AGRICULTURE AND FOOD SAFETY

AGRICULTURE AND FOOD SAFETY

Editor
Ramniwas Sharma

BIOTECH
2014
Biotech Books
Delhi - 110 035

ISBN 81-7622-163-5

Published by : **BIOTECH BOOKS**
1123/74, Tri Nagar,
DELHI – 110 035
Phone: 27382765
e-mail: biotechbooks@yahoo.co.in

Showroom : 4762-63/23, Ansari Road, Darya Ganj,
NEW DELHI - 110 002
Phone: 23245578, 23244987

Printed at : Tarun Offset Printers, Delh

PREFACE

Food science is a discipline concerned with all aspects of food—beginning after harvesting, and ending with consumption by the consumer. It is considered one of the agricultural sciences, and it is a field which is entirely distinct from the field of nutrition.

The book is revealing how modern technology has drastically affected our environmentally based, technology enhanced food supply. Exposing myths and presenting extensive, meticulous research on the history of pesticides, pollution, organic agriculture, and much more. It is a very highly recommended, fact-filled primer, which will provide the non-specialist general reader an invaluable instructional background.

It is the ultimate irony that the anti-technologists, who oppose irradiated, genetically altered, and biotechnological foods, are harming the very people—whom they blatantly otherwise claim to defend—who most need the potential bounty of that advanced nutrition.

Arguing that humans are inherently technological beings and that technology has supported immense human progress over the past century. Overall, the book is a valuable resource for its soup-to-nuts information approach. The book, a gold mine of useful information for post graduate students and researchers of relevant disciplines and also it should hold the attention of those having in recent trends and advances about the world of food safety.

Editor

CONTENTS

CHAPTER 1

CHALLENGES AND OPPORTUNITIES OF FOOD SYSTEM

A few generations ago, most of population lived on farms or in small towns. Nearly all food was grown at home, or nearby, and processed by own families or by people. The technologies were simple and available to all. The remainder of that picture, however, was not so comfortable. Frozen foods, iodized salt, vitamins, enriched bread, and air transport of fresh foods were unknown.

For all except the very wealthy, fresh fruits and vegetables were limited to what was seasonably available. Goiter, rickets, beri-beri, and pellagra were common. Today, nutrient deficiency diseases in the Western world are a distant memory. A huge variety of food is available year round. For this to be possible, many of these foods are grown thousands of miles away from where we live, and processed by people we neither see nor know.

Furthermore, in the United States, expenditures for food are among the lowest in the world-about 10% of average family income. Supporting those facts is an enormous breadth of science and technology. That technology is no longer simple and familiar to all. It is complex, and to most consumers, unknown. Discussing the benefits and concerns that biotechnology creates requires discussing these usually unfamiliar technologies into which biotechnology fits.

History teaches another aspect that must be addressed in the course of introducing any new technology. Except for some life-saving medical advances, and sometimes not even then, it is rare for a new technology to receive a broad and enthusiastic welcome. Canned food, for its first hundred years, was viewed apprehensively, and not without reason. In those pre-bacteriology days, it was far more an uncertain art than a solid science.

Pasteurized milk, a life-saving technology in its elimination of the microorganisms causing tuberculosis and undulant fever, was originally viewed with deep suspicion. Artificial insemination of farm animals -critical in selective breeding of improved livestock-was regarded as tampering with nature. Margarine was opposed, partly for alleged health concerns, but mostly because it was a threat to the dairy industry. All sorts of health threats-far beyond pacemaker interference-were originally attributed to microwave cooking.

Such apprehensions were by no means confined to food, for which we have always understandably felt a close personal concern. These examples, a few among thousands, illustrate the mix of motives-some rational, some not, some economic, some religious or ethical, some based only on lack of understanding-that have characteristically attended the implementation of innovation. Biotechnology is no exception.

There are numerous specific benefits of recombinant DNA biotechnology-derived foods. With the use of rDNA biotechnology, there is the potential for enhancing plant availability and growth, and the ability to grow more and better food with increased nutritional value, including improved animal feed. rDNA biotechnology is expected to revolutionize food bioprocessing through improvements in the responsible microorganisms (e.g., bacteria, yeasts, and molds) and efficient production of specialized enzymes and ingredients via fermentation technology.

rDNA biotechnology also creates the opportunity to produce edible vaccines and therapeutics for preventing and treating diseases. The use of rDNA biotechnology also has specific environmental benefits, with the development of new varieties of crops that exhibit increased resistance to pests, tolerance to more environmentally benign herbicides, and virus resistance.

Nearly all productive land is already growing food and water resources accessible to agriculture are, if anything, shrinking. Intensification of production emerges as the most promising option. The emergence of biotechnology has both been driven by and fueled pesticide and seed industry consolidation. Across the food chain, companies are merging and industrylines are blurring. On both the input and food processing and marketing sides, the U.S. food system is rapidly becoming more concentrated and integrated. Despite the rate of change in production agriculture, the scope and pace of structural change in the basic production sector is clearly much less than in the input and processing/marketing sectors.

SUSTAINABLE AGRICULTURE AND FOOD SYSTEM

Sustainable agricultural and food systems must -

— Provide a reasonable rate of return to farmers to sustain farm families, agricultural infrastructure, and rural communities;

— Assure a reasonable rate of return to public and private providers of farm inputs, information, services and technologies;

— Meet society's food needs;

— Preserve and regenerate soil, water and biological resources upon which production depends, and avoid adverse impacts on the environment;

— Increase productivity and yields at least in step with growth in demand; and

— Adhere to social norms and expectations in terms of fairness and equity, regulations, food safety, and ethical treatment of workers, animals and other creatures sharing agricultural landscapes.

Agricultural systems and technology become unsustainable for a variety of reasons, but loss of profitability is almost always the most immediate. When an agricultural system or technology becomes unsustainable, agriculture evolves through changes in ownership or management, or by an infusion of technology, capital or new enterprises. It has done so for decades, and in some countries, for centuries.

The only thing that really changes is that energetic and ambitious managers willing to accept lower returns per bushel find the capital to expand, maintaining their income only by expanding their acreage base. One of the tragic dimensions of recent changes in U.S. agriculture is that forces beyond the control of producers can drive even the best operators, with generations of hard-won equity in a farm, off the land. Hundreds if not thousands of Illinois hog producers may not make it through the winter. On conventional and sustainable farms alike since the 1980s, the slide toward unsustainability has typically been triggered by excessive optimism and misguided investments when prices are high, or changes in market conditions or policy that trigger significant price swings.

In some regions the ravages of bad weather and flooding, or a decision to close a processing plant or livestock auction, have taken a toll. But now another generic pitfall is emerging-poor judgement in the selection and use of agricultural technologies. For two reasons this new source of instability may rise in prominence in the years ahead. First, increasingly serious economics surprises and setbacks. Many emerging biotechnologies are more expensive to bring to market. The costs of creating and protecting intellectual property are already high and bound to rise, as are the steps required in overcoming regulatory hurdles.

Many new technologies cost more per acre at the grower level and require more management, information, and skill to use to best advantage. When they fail, they tend do so more profoundly than past technologies, imposing greater losses on growers. And some biotechnologies will constrain marketing options - the last thing farmers need as input suppliers, the grain trade, and processors compete to gain a larger share of the consumer food dollar, often at the expense of farmlevel profit margins.

Second, a growing array of sources of genetic, ecological and biological instability. Both the nature of many new technologies and how they are being marketed are setting the ecological foundation for trouble. The pesticide treadmill is giving way to the technology package treadmill. But now, instead of just destabilizing the interactions of pests, beneficials and crops, technology packages are throwing crop genetics and physiology into the mix, as well as the potential for a new generation of highly mobile ecological problems.

Instances of erratic performance will no doubt continue to plague some transgenic varieties. Problems arising in areas where transgenics are used will occasionally move onto surrounding farms where the technology may not have been used. Sorting out who caused what, the magnitude of losses and who must pay will be costly and contentious. These costs will be factored into per acre expenditures in one way or another.

Who will bear the economic burden of these costs and risks remains unclear, but recent developments suggest most companies will be aggressive, arguing that problems arose because of bad weather or management shortcomings. Hence, farmers are likely to bear the burden of proving harm and causality, a burden not always easy to sustain near the frontiers of agricultural biotechnology, as some farmers are learning.

Global Food Needs

According to the FAO over two billion people suffer from

deficiencies in micronutrients and close to a billion people remain undernourished, yet farmers worldwide struggle with low prices caused by over-production. Dietary deficiencies and hunger persist because of poverty, politics, and the design of food systems, not because of a lack of production. In the developing world natural resource degradation continues to worsen as more and more people try to survive on less and less land. The number of people living in harm's way-flood plains, beneath or on steep hillsides, in arid regions-is growing, exposing countries and ecosystems to new sorts of natural calamity.

American agriculture has never been comfortable with nor rewarded those inclined to think outside the box, whether in academia or industry. Most scientific and commercial breakthroughs have come from specialization and reductionism. The day may come when limited thinking will begin to undermine the inclination to innovate that runs so strong in American agriculture. There are many paths to intensification in agriculture yet only a few receive meaningful institutional and financial support. Different paths are likely to lead to different outcomes.

No one path will work everywhere nor meet all needs. For sound ecological and economic reasons, the nation will benefit from multiple agricultural intensification choices and the knowledge and ability to choose wisely among them. Finding how to do so is one of the land grant system's most daunting contemporary challenges.

TECHNOLOGY AND SUSTAINABLE AGRICULTURE

Some farmers sense that they have climbed on a treadmill and that the harder they run, the faster they must go to just stay even. Lloyd Fear notes that the more technology helps farmers increase production, the softer prices become, especially as governments pull out of the supply management business in deference to the need of the grain trade for maximum volumes and unconstrained price movement.

Among the implications of changes underway in farming, Lloyd Fear believes that the "farmer is in an industry where his fortunes and future are being dictated by others with little thought to the impact of their actions." So tonight let's do a "little thinking" about technology, what is driving it, and implications for farmers, consumers, and the country. Will make five major points -

— Biotechnology is fundamentally changing the nature of technological change, and in due course may reconfigure the economic, cultural and ecological genome of the farm community.

— Past agricultural science and technology revolutions have been largely public sector financed and driven, and for the most part, responsive to farmer-needs. The biotechnology revolution is private sector and corporate profit-driven.

— For at least another generation the real prices of most farm commodities are likely to remain flat, or at best rise modestly. Input industry and processing sector growth, debt-service and profits will have to come from somewhere. Expect further pressure on farm profits.

— Biotechnology is driving change at all levels of agriculture and at an unprecedented rate. The law of unintended consequences is likely to have many field days. The pace of change will bring upon rural America and our food system selection pressures unlike any in the past.

— If current trends continue, farmers and society may be able to do relatively little in shaping these changes, the genie is out of the bottle.

Impacts of Biotechnology

Methods to project the various impacts of emerging biotechnologies are needed for many purposes. Assessments of biotechnology should evolve from the principles and concepts of ecology and evolutionary biology rather than mechanistic, input-output, dose-response models. Conventional approaches to risk assessment pay little heed to adaptive forces and evolutionary biology. This needs to change since synergism between plant health, stress, genetic adaptation, and changes in biodiversity will trigger many unintended consequences.

It is also important to question who will benefit from various applications of biotechnology. Clearly the pesticide and seed industries are bullish on biotechnology, a reflection of expectations that they will benefit from the opportunity to offer farmers genetically engineered seeds and production inputs. Many academic scientists working on molecular genetics are also clearly eager to apply their skills in creating new plants.

In fact, all the public and private institutions of agriculture, both the USDA and EPA, the farm press and the private sector have, for all intents and purposes, embraced biotechnology as the wave of the future. For better or worse the row crop farmers of the Midwest are now riding that wave.

Roots of Technology

Each generation of farmers has to deal with a set of yield, production and profit constraints. These include emerging and well-established pests, and maintaining and hopefully building soil quality. Illinois farming systems have grown more productive per acre through integration, over many years, of a series of new technologies. Farmers will become more productive and prosperous in the years ahead if they are able to wisely choose among expanding technological options and are not overwhelmed by anti-competitive forces in the marketplace.

Farmers make a series of decisions each year that determine what will be grown, where and how. These decisions and how skillfully they are carried out join the weather and market prices in determining income from farm operations. They include -

— *Farming Enterprise Design*: (1) whether to specialize in crops, livestock, or manage a mixed system; and (2) mix of crops and crop rotations, hence defining the farm's mix of marketable commodities.

— *Selection of Genetics*: complex tradeoffs arise in choosing between high yield potential, resistance to pests and stress, and output traits. The emergence of GMOs requires farmers to weigh new factors and tradeoffs in choosing varieties.

— *Agronomic Systems*: choices span tillage and planting systems, fertilization methods, weed and other pest management practices, soil and water conservation practices, and harvest timing and methods. Some decisions -adoption of no-till - forego certain options and heighten reliance on certain technologies and inputs.

Some decisions have long-run consequences - whether to diversity into livestock, or specialize in grain production, the purchase of no-till planting equipment or a second combine. Other decisions must be made quickly as a season unfolds and may, or may not improve profits in a given year. Examples include whether a second pass is needed with a cultivator or how much nitrogen to apply.

Technological innovation - the kind farmers welcome and society has been willing to support-occurs when a farmer is able to make these interconnected decisions in a way which, on average over several years, increases yields more so than costs, or reduces costs more so than yields, while either improving or leaving unchanged the environmental and resource base impacts of farming operations. In the last two decades there has been relatively little change in many aspects of farming system design in Illinois.

Corn and soybeans dominate the landscape and have for years. Most farms specialize in crop or livestock production and the trend toward large-scale operations seems, if anything, to be gaining momentum. The effort to find a viable third crop for corn-soybean rotations is no one's priority over the long haul, to the development of new crops and more diverse rotations and farm enterprises, a task only public sector investment can tackle.

Accordingly, innovation in farm system design appears not likely to be a major source of change, despite great potential to improve the performance of Midwestern farms. A third crop might largely solve the problem with several corn insects. During the 20th century most major technological changes have emerged from research carried out within the land grant university system and USDA in response to natural resource, farming system, genetic or biological constraints.

The private sector always played a vital supporting role in making new technologies commercially viable and delivering them to farmers. The polarity has clearly shifted since the early 1980s. Private companies have always and will always work to make a profit, as farmers also work to earn profits. Profits keep the system vibrant and investment capital flowing into agriculture. But new issues are arising because technology and industrial restructuring are fundamentally shifting the "terms of trade," and hence the relative economic performance and power across sectors in the food system.

In the past, seed, farm machinery and pesticide manufacturers depended directly on a healthy farm sector to assure that their customers would return to the marketplace every year, and agribusiness supported policies that helped keep farmers in business and eager to increase productivity. But now, across many fronts, input suppliers, processors and integrators are getting more directly involved in the management and control of farms, and blurring the line between farm and agribusiness income, profit, decision-making, and risks.

In addition, a growing share of the research effort in the land grant system is problem or crisis driven and short-term, quick-fix oriented, as opposed to focusing on discovering the genetic or biological basis for inherently more productive and profitable farming systems over the long-run. The near-term problems demanding attention often arise from the combination of chemical technologies and homogenous, large-scale production systems. In the world of biotechnology, the land grant system is now struggling with a dual role.

The system has been and remains a major developer and promoter of biotechnology. But second, it is also the only public institution with the expertise and ability to project and understand the myriad potential adverse ecological, biological and economic consequences of biotechnology.

HERBICIDE TOLERANT

Roundup Ready soybeans, other herbicide-tolerant varieties and Bt-transgenics clearly comprise the first wave of the biotechnology revolution. After just two years of widespread commercial use, over one-third of U.S. soybean acreage will be planted to Roundup Ready varieties in crop season 1999 - a remarkable and unprecedented change in weed management system technology. Given how fast Roundup Ready beans have gained marketshare, one might infer that farmers have recently been plagued by serious problems managing weeds in soybeans.

But actually in the 1980s and the 1990s the range of soybean weed management systems and technologies has exploded. Ridge tillage, no-till, banding, improved cultivators, newly registered postemergent herbicides and new planting systems gave farmers many new options. The chemical toolbox is overflowing - more than two-dozen new active ingredients have been registered in several families of chemistry. The first herbicide-tolerant soybeans were planted on a commercial scale in 1996.

The problem for farmers seeking herbicide-based solutions has become the complexity of choices and compatibility with soil types, varieties, and farming systems, along with containing cash costs. The mid-1980s Monsanto business managers recognized in biotechnology a way to extend high profit margins from Roundup sales beyond the end of patent protection in the year 2000. They knew the price of Roundup would come down as other manufacturers entered the market, a process that has already begun.

But by developing and marketing Roundup Ready plant varieties, Monsanto could, in effect, transfer high profits from sale of the chemical to "seed plus technology packages" accompanied by a technology fee. Indeed, the announcement a few weeks ago that Monsanto is lowering the price of Roundup but increasing the Roundup Ready technology fee is consistent with such an overall plan.

Weed Management Systems

Three criteria should be applied in judging weed management systems and technology

— Long-term Effectiveness-the scope and difficulty of weed management challenges over time.

— The goal should be lower weed seed density and fewer outbreaks over time.

— Robustness-the ability to respond to unexpected conditions and shifts in weed pressure.

— The goal is more and better tools to deal with whatever problems emerge; in short, greater system resiliency.

— Weed Management System Costs.

— The goal is lower weed management costs per bushel produced and as a percent of net crop income.

Applying these criteria in evaluating Roundup Ready beans is a reasonably straightforward exercise. When first introduced, Monsanto promised that Roundup Ready soybeans would greatly simplify weed management. Just plant, spray once and relax for the rest of the season. If the technology had lived up to this early billing, it would have satisfied the first two criteria. But in the field, nature's proclivity to adapt has made things much more complicated and the future uncertain for users of this technology.

Use of an over-the-top non-residual systemic herbicide like Roundup will generally work well in managing the weeds that have recently germinated, but will miss weeds that germinate either earlier or later. Growers planting RR soybeans have, as a result, three choices -

— Apply early to avoid early-season yield losses from fast growing grasses, possibly suffering yield losses from late season weeds;

— Delay applications until most weeds have germinated, risking loss of yield to early weeds; or,

— Apply Roundup twice or more, and/or a residual herbicide at planting or when Roundup is applied.

Most are choosing the third option, since there is a very low tolerance among growers for either weeds in fields or yield losses. The emergence of tolerant and/or resistant weeds is a second mechanism likely to undermine the long-term effectiveness of Roundup Ready soybean systems. Indeed it is already happening. Tolerant populations of waterhemp, sometimes called pigweed, are emerging across the Corn Belt.

This particular species of weed is highlighted in the 1998 edition of the Weed Control Manual as the most worrisome "Up and Coming Weed" both nationwide and in the North Central region. Reasons why include the emergence of resistance to triazine and ALS herbicides and its ability to withstand Roundup applications. Because of resistance and weed shifts, Roundup Ready soybeans clearly flunk the first criterion since weed management problems are already getting worse on farms relying on Roundup Ready technology.

Much evidence, ecological theory, and practical pest management experience suggests such problems will worsen on farms that continue to rely mostly on Roundup Ready varieties. The combination of Roundup Ready beans and Roundup Ready corn in a rotation will further torque the ecology of weed populations and will eventually bring on the loss of commercial viability of this technology. The second criterion involves the robustness of weed management systems -- the capacity of farmers to respond to unexpected circumstances, whether brought on by weather, the evolution of a new weed biotype, or a change in tillage systems.

Roundup Ready systems get a passing grade because the systems do not force farmers to forego most other weed management options. Farmers can still choose to apply other herbicides. Early season cultivation remains an option, as does diversifying rotations and use of cover crops. Cost is the final criterion and the one most likely to limit adoption. Coverage in the farm press makes it clear that Roundup Ready systems offer farmers modest potential to reducze costs per bushel IF everything works well - i.e., there is no yield drag and one application of Roundup is sufficient in managing weeds.

When a second application of Roundup is added, or another herbicide, the economic advantage largely disappears. In parts of the Midwest the costs of Roundup Ready soybean systems have taken off. At a recent field day the Adair County Soybean Association released its projection of the costs of a "total" Roundup Ready system in 1999,

— Burndown with Prowl, 2,4,-D and Roundup Ultra, $19.40 per acre.

— June spray with 2 pints/acre Roundup Ultra and adjuvants, $15.31.

— July respray with 1 pint Roundup Ultra, $9.06.

— Seed cost per acre (includes $5.00 technology fee), $25.00.

Many farmers have been willing to accept such increases in costs in return for the perceived advantages in managing weeds. But with crop prices heading down, several new herbicides gaining registration, and growing evidence of a yield drag, farmers will begin to look elsewhere for weed management options. Monsanto is already taking steps to preserve its marketshare by lowering the cost of technology, and further price reductions can be expected in the future.

Seed plus herbicide costs are bound to rise further. Everyone expects that a higher percentage of soybean acres will be planted to GMO varieties and that most farmers doing so will need to apply two or three applications of two or more herbicides. Weed shifts, resistance and aggressive marketing programs will work together in unpredictable ways driving further change in weed management systems. For a variety of reasons, the costs of seed corn plus weed management are also likely to rise sharply in the next few years.

Consider the implications of one justannounced corn program, sure to serve as a model for future soybean programs. AgrEvo and Novartis are offering a "guaranteed" corn weed control for farmers planting AgrEvo's Liberty-Link corn who also buy and apply Novartis' Dual Magnum II benoxacor) or Bicep II (S-metalochlor + benoxacor + atrazine) herbicide. That's up to four active ingredients in managing weeds after planting a variety that supposedly makes weed control easier. Offering guarantees as a marketing ploy has been strongly criticized in the past for establishing unreasonable expectations among farmers, ratcheting up cash expenditures, and encouraging excessive herbicide use, which can trigger resistance and other problems.

Despite some progress in recent years in constraining the scope and expense of herbicide guarantee programs, some companies envious of Monsanto's successes with Roundup Ready varieties are bound to offer farmers a range of incentives to switch to their "technology package. " Higher weed management costs, weed shifts and resistance are not the only problem that may evolve from widespread planting of herbicide-tolerant crops. Changes in soil microbial communities caused by Roundup and other low-dose herbicides can both increase the vulnerability of beans to Pythium and reduce the uptake of phosphorous.

INSECT PEST MANAGEMENT

New insect and weed management problems in the corn-soybean rotation arise through one of several mechanisms from selection pressures inherent in common rowcrop agricultural systems. Each new challenge reflects the success of adaptive forces that will, in the years ahead, work to undermine the effectiveness of emerging technologies. Cost-effective use of pest management technology, regardless of its genesis, depends upon the degree to which it helps diversify and complicate the challenges faced by pest species within farm fields.

Many technologies once heralded as major innovations have failed because of agriculture's tendency to rely on technology to simplify and homogenize systems rather than to diversify them. To correct the underlying problem, farmers need to look at their farms and fields through the lens of evolutionary biology. Farmers will either learn to manage selection pressure or will continue to be managed by it. Turning selection pressure into a positive force will pose many new challenges for land grant scientists and farmer-networks.

A new generation of diagnostic tools are needed. The chemical cues governing soil microbial interactions with roots and pathogens will need to be isolated, along with threshold levels that seem associated with disease suppressive or conducive soils. Others will then need to work on the discovery of genetic, biopesticide and cultural practices that can help trigger resistance mechanisms sooner and more effectively.

Evolutionary Biology

The most important recent change in insect pressure in the Midwest has arisen from the behavioral adaptation of the Western corn rootworm (WCR) to the longeffective corn-soybean rotation. This simple, moneymaking rotation has been one of the great success stories in the world of cultural pest management practices, reducing corn insecticide use by half, lessening water quality problems and reducing grower cash costs.

On continuous corn in the 1990s, 90 percent or more of corn acres are treated with a soil insecticide while less than 15 percent of rotated corn has been treated. The success of the corn-soybean rotation in limiting WCR losses set the stage for trouble, as farmers and scientists were lulled into thinking that little more was needed to be done in dealing with this pest. Economically significant Western corn rootworm damage in first-year corn following soybeans was first documented in isolated fields in the mid-1980s.

Over the next decade the pattern of infestation was uneven and episodic. Some seasons, like the wet crop year in 1998, do not favor WCRs and hence mask changes in pest behavior and damage potential. But over time it has become clear that in many parts of the Midwest, the Western corn rootworm had developed what scientists call behavioral resistance to a management-based control strategy, in this case crop rotation.

Western corn rootworm adaptation came on gradually. Its roots were subtle, hard to isolate from other dynamic factors affecting pest complexes. Both farmers and entomologists did not realize the significance of their observations of WCR damage in first year corn until the mid-1990s. In 1993 some farms in east-central Illinois suffered serious WCR larval injury in first year corn root systems. In both 1995 and 1997, damage was "severe and prevalent" in east-central Illinois,

Field research quickly ruled out extended diapause and repellency from use of synthetic pyrethroid insecticides as possible causes, leading to the realization that the rotation itself had broken down as an effective control strategy. The narrowness of the control strategy used against the rootworm explains, at least in part, its ability to adapt around it. If other control measures had helped spread out the control burden-lessening the selection pressure, in effect managing resistance- farmers might not be facing this new problem today. There is an important lesson embedded in the events leading to WCR adaptation:

First principle that farmers and scientists should heed in shaping pest management systems and fitting biotechnologies within them. The best way to preserve an effective and affordable pest management technology is to use it within a diversified system that spreads the annual control burden across differing mixes of cultural, genetic, chemical and biological tools and tactics.

Collective experience gained in managing hundreds of pests across thousands of cropping systems leads to some general hypothesis. While the number of tactics needed for stable control will vary across crops, pests and climatic conditions, four general hypotheses can be stated in simple terms as -

— A pest management system that relies predominantly on one tactic is inherently vulnerable. The more generations of a pest in a season, and the longer the pest is subject to selection pressure from a single tactic, the greater the odds the pest will adapt around the tactic and adapt quickly.

— Systems that spread control in a given crop cycle over two substantially different tactics remain vulnerable, but to a lesser degree. Two unrelated tactics used together in a given year will be more effective, in most circumstances, than two tactics used in isolation, one the first crop cycle, another during the second. Two different sets of two unrelated tactics in successive crop cycles will further tip the odds in favor of farmers and sustainability, and may be essential in the case of insects with multiple generations per season.

— Systems with three or more distinct but significant control tactics are likely to be sustainable in the hands of good managers who act upon the first signs of trouble by further diversifying and complicating the tactics mixed within integrated control systems. By "significant," mean a control tactic that in some years can be effective alone and which rarely fails when combined with another major tactic.

— In all pest management systems, there is a minimal degree of redundancy in control tactics below which the odds of adaptation steadily improve toward certainty, as well as a degree of redundancy above which adaptation becomes unlikely and easily managed if it begins to appear.

Testing these hypotheses and establishing the thresholds noted in the fourth hypothesis should become an important focus of land grant pest management professionals. Imagine how helpful such information would have been in 1986 when problems first arose with the Western corn rootworm in first-year corn. If Illinois entomologists had known then that insects like the WCR are likely to adapt around a cultural control tactic when it is the sole tactic used, they might have realized more quickly what was happening in the field and how to avoid it becoming a stable, statewide problem.

Corn Insects with Transgenics

The other big changes in corn insect pest management are occurring as a result of the introduction of Bt-transgenic varieties that overcome attack by second generation European corn borer. When first introduced, Monsanto and other companies marketing Bt-corn varieties acknowledged the possibility of resistance but argued that the high-dose expression of the Bt toxin, coupled with a 5 percent refugia planted to traditional varieties, would preclude the emergence of resistance.

New science and field experience have confirmed the lack of a solid science base supporting key assumptions that underlie the Bt-transgenic resistance management plans (RMPs) accompanying transgenic crops released to date:

— Bt endotoxin expression under field conditions is not uniform across fields or within plants, providing some insects with a chance to avoid plant tissues delivering a lethal dose. Plus Bt expression tapers off as the season progresses and any source of stress or plant injury can lead to differential toxin levels and some non-lethal foliage, which insects seem able to find.

— Two studies published in 1997 show that resistance to Bt is not as rare as previously suspected. Dr. Gould and colleagues reported that the frequency of the major Bt resistance gene in a field population of Heliothis virescens was about 0.0015. Dr. Tabashnik and colleagues showed that the frequency of a multipletoxin resistance allele in susceptible populations of the diamondback moth was an astonishing 0.120, evidence that resistance gene carries very little, if any, "genetic load".

— Resistance will emerge quickly to multiple strains of Bt endotoxin since the resistance gene appears to be a dominant trait. Tabasnik and colleagues found that diamondback moths "share a genetic locus at which a recessive mutation associated with reduced toxin binding confers extremely high resistance to four Bt toxins". Furthermore, data from a Florida population of diamondback moth demonstrated similar results. At the 1997 Entomological Society meetings in Nashville, TN, Dr. Tabashnik stated that he expected that findings similar to his would emerge in ongoing work with several other lepidopteran species.

— Work by Dr. Angelika Hilbeck of the Swiss Federal Research Station for Agroecology and Agriculture has shown that Bt-corn can adversely impact populations of key beneficial insects. In a number of trials, Dr. Hilbeck reported 60 to 65 percent mortality among lacewing larvae that fed on lepidopteran larvae reared on Bt corn.

Resistance to Bt

Conventional wisdom has changed rapidly in the world of Bt resistance management. Just a few years ago the companies introducing Bt-transgenic seeds downplayed the prospect of resistance and submitted to the EPA reams of documentation and modeling that supported the efficacy of the high-dose plus refugia strategy. Most university experts had little to contribute as EPA reviewed Monsanto's Btcorn resistance management plan, since they had not been granted access to any seed or detailed information on the technology until the year commercial introduction began.

The few that had been granted access to seed or had helped develop the technology were constrained by secrecy agreements and/or conflicts of interest. Development, testing, and wide scale introduction of a major new technology in near-total secrecy is a new development that undermines some of the corrective "feedback" loops that have helped make our system of technology development reasonably reliable.

Hearing no science-based objections from the public, EPA assumed that no one had uncovered any problems and that hence, problems were not likely to emerge. The veil of secrecy now surrounding agricultural technology development efforts raises an important public policy issue. EPA regulations and rule-makings depend heavily on informed and disinterested public comment to assure thoroughness and balance, since EPA cannot maintain "critical mass" in all areas of commerce it is responsible for regulating.

When scientists are unwilling to share data, are constrained in what they can report, and/or have no opportunity to study new technology, public institutions and regulators have to fly blind for a period of time. Beginning in about 1993 a small number of public interest groups started to voice strong concerns about Bt-transgenics. In 1996 the Consumer Union (CU) book Pest Management at the Crossroads set forth the arguments why

EPA should withdraw the conditional registrations granted to Bt-transgenics and called upon the agency to place a moratorium on any further approvals until proven, science-based and enforceable resistance management plans were in place.

Since release of the book, CU has submitted several technical assessments on the risks associated with Bt-transgenics . As new Bt-varieties reached farmers, a growing number of academic scientists startessd to assess the likelihood of resistance. Evidence mounted that the concerns were real and that decisive steps were needed, and soon, to preserve the effectiveness of Bt against a wide range of key Lepidopteran insects. Just in recent weeks, Monsanto and other companies have agreed to the need for at least a 25 percent refugia in the Corn Belt, and a 50 percent refugia requirement in the south, where Bt-corn and cotton are often grown in close proximity.

While details remain sketchy on these plans, this unexpected decision by Monsanto is evidence that they feel there is, in fact, a cause for preventive action. Other companies are being even more direct. An incentive program has just been announced by Novartis. Growers that buy the appropriate share of Bt- and non-Bt seeds will qualify for a cash incentive. More such offers are bound to follow.

ROLE OF GENETICS AND BREEDING

Every year, Pioneer maize researchers around the world evaluate about 130,000 new experimental hybrids. These hybrids enter a four to five-generation testing cycle...The top 10 percent - 13,000 - . And finally, from about 130,000 original candidates, only about 15 to 20 hybrids 'graduate' to commercial status. The odds of a hybrid reaching the farmer have accordingly dropped from about one in 3,000 in 1979 to one in 7,500 today.

As the science and art of genomics develops, it is likely the number of hybrids screened will increase and the odds of success will continue to decline, although the scope of genetic assessment will surely expand and in time the efficiency of breeding efforts will improve. Since 1970 Pioneer Hi-Bred's investment in corn breeding has risen steadily from less than $10 million per year supporting some 150 scientists to about $150 million in the 1990s, supporting 550 scientist years. Accordingly it is clear that it is taking more and more effort to sustain the 1.5-bushel average yield gain per acre that breeders have historically delivered.

The performance of 36 widely grown and successful hybrids released between 1934 and 1991, including some open pollinated varieties. The genetic yield potential of maize hybrids has increased 74 kg/ha/year, a little over 1 bushel per acre, according to a large number of trials carried out in 1991-1994. According to Duvick, "Maximum yield potential per plant has neither increased nor decreased during the past 70 years, as measured on non-stressed plants grown at very low densities."

Cassman and Duvick cite other evidence to support the conclusion that the yield potential of hybrids under optimal conditions has not risen for decades. For example, the top irrigated corn yields achieved by contest winners in Nebraska plateaued at just over 300 bushels per acre decades ago and have changed little since. Accordingly, the ability to perform well at high plant populations per acre accounts for most of the corn yield increases since the 1940s.

Over time higher yielding varieties were improved in their ability to resist root lodging, premature death and stalk rot, second generation European corn borer, and other stresses associated with high density planting . Several other traits have not changed over time - plant height, grain moisture at harvest, resistance to first generation European corn borer, and grain percent oil. Duvick also points out that breeders have pushed yields up about as much as possible through the manipulation of a number of traits including tassel size, grain protein percent, and upright leaf habit.

Consolidation of Seed and Pesticide Industry

For decades the private sector has dominated breeding activity in crops like maize that benefit from crossbreeding. In the case of corn and most hybrids, achieving higher yields reliably has been the dominant goal driving private sector plant breeding. Traditionally, the public sector was largely responsible for improvement of selfpollinating crops, where the focus has often been resistance to diseases and insects.

With passage of plant variety protection laws in the 1970s, commercial breeding activity has increased. According to Duvick, the increase in private funding for breeding has triggered declines in public sector support "almost in inverse ratio to the increase in private activity". The integration of the seed and pesticide industries has been driven predominantly by the opportunity to use genetic engineering to preserve profit margins for proprietary pest management technology.

There is some evidence already of one impact of changes in corporate structure- seed plus pesticide technology packages appear to be eroding the traditionally high rate of return farmers have enjoyed from investments in improved hybrids. For decades the rule of thumb has been that the seed industry has delivered $3.00 of added income for every additional dollar spent on hybrid seed. Corn cost of production, yield, and price data comes close to supporting this general rule of thumb over the period 1975 through crop season 1997.

Corn and Soybean Seed Prices

First, the integration of the seed and pesticide industries has been costly. Competition for seed company assets across the historically larger and more profitable pesticide industry has driven up the value of seed companies by at least 50 percent in a matter of a few years. Inflated stocks have inflated appetites for capital and income. Accordingly, industry integration has required a greater share of grower-income to service debt and deliver an acceptable rate of return to shareholders.

In addition, in the last few years seed companies have increased the portion of income devoted to research from the historic industry average of 5 percent to 8 percent of sales to 12 percent to 15 percent. Since the seed industry used to operate at much lower profit margins than the pesticide industry, financing this increase in R+D expenditures has placed upward pressure on prices on seed and/or technology fees. Second, since the mid-1990s a growing portion of the focus of breeders has been devoted to the introduction of either herbicide tolerant or Bt genes into elite germplasm and top-selling hybrids.

This has increased the costs of developing, patenting, obtaining the rights to use, and defending intellectual property. For each seed company and across the industry as a whole, there are no publicly available data to settle the issue of whether the work required to introduce herbicide tolerance and Bt genes was, in effect, added on top of ongoing efforts to increase yields and productivity, or whether there were tradeoffs. To the extent the attention of breeders shifted, there will likely be lesser annual yield gains for the foreseeable future, at least compared to what they otherwise might have been.

Plant genetic enhancement is a deliberate process requiring careful empirical study of thousands of crosses in tens of thousands experimental plots and fields trials. The rush to market genetically engineered seeds may have compelled some seed companies to cut corners and market varieties before breeders had adequate time to assure all necessary traits were fully expressed, such as resistance to common pathogens. With breeders focusing on traits related to pest management and herbicide compatibility, they were not able to pursue other crosses with greater potential to increase yield.

Third, pesticide industry profit expectations and margins are different from the seed industry. In order to win and hold customers, the hybrid seed industry has tried to assure that the farmer earns, on average, $3.00 for every extra dollar spent on seed. In order to keep yields rising and grower returns high, the seed industry has had to operate on about 12 percent to 15 percent margins, and has invested about half of its profits back into research and field breeding activities. For example, in 1987, Pioneer Hi-Bred spent $49.9 million on research, or 5.9 percent of gross sales and over 50 percent of operating profits.

The pesticide industry, on the other hand, invests more heavily in research and incurs much higher regulatory compliance costs. Manufacturers often earn 30 percent to 50 percent profit margins on sales of proprietary chemicals. The herbicide Roundup (glyphosate) is a notable example. This one product has accounted for over half Monsanto's total profits in some years, and has generated income on the order of $300 million plus on sales over $1 billion.

Net income from herbicides, mostly alachlor (Lasso) and glyphosate (Roundup), accounted for 59 percent to 85 percent of Monsanto's net income in 1981-1983, but represented only 18 to 21 percent of total corporate sales. Even after supporting R+D expenditures falling between 12 to 16 percent of sales, net returns of 10 to 15 percent of sales are common in the pesticide industry. Operating profits on the order of 20 to 30 percent - at least double those common in the seed industry during the 1970s through the mid-1990s. Clearly, if pesticide manufacturers priced their new products the way Pioneer prices seed, the economics of row-crop agriculture would be more favorable to farmers.

Yet the profit expectations of the pesticide industry appear poised to dominate the marketing and pricing decisions covering the sale of seed and pesticides. If this trend continues, input suppliers will wind up with a greater share of the profits associated with the production of a bushel of corn or soybeans.

Plant Breeding

In most instances, companies will have the opportunity to choose between developing new genetic or chemical solutions to emerging, or old pest problems. In many instances, they will offer farmers both resistant varieties and pesticides. In some cases applications of proprietary chemicals will be required to trigger or otherwise take advantage of novel genetic traits introduced into GMO plants. Both types of technologies will make farmers more dependent on other sectors of the food system, on the one hand for inputs and pest management technology, and on the other in the marketing arena.

Both these trends are likely to enhance the economic power and profitability of the input and processing sectors at the expense of farmers. This should, in turn, attract more capital and investment in these sectors - which historically has been a good thing. But as stated earlier, there are new concerns. The U.S. food system is already highly integrated and becoming more so. The performance of the whole system depends on the capacity of each sector to respond to challenges and market opportunities, and to innovate.

The chronically poor economic returns to the farm production sector, wide swings in prices, profits, and debt load, the backing off of public sector investment in science critical to farmers, and disinvestment in farm level natural resource management all suggest that the farm production sector is emerging as the weakest link in the system. To some the logical solution is for stronger parts of the system to simply take over production responsibilities. This is, indeed what appears to be happening in some parts of agriculture.

Farmers need the help of agricultural economists in understanding the longer-term consequences of these trends, especially any changes that limit their access to markets or the role of open competition within markets in setting prices. There are other reasons to worry about changes in the seed industry. Cases may emerge where companies forego the chance to breed resistance into a variety since higher profits can be earned by continued sale of a pesticide. While breeders have focused single-mindedly on increasing yields in the past, in the future their priorities are likely to be shaped at least to some degree by the need to create synergy between a company's pesticide and seed products portfolio.

The willingness of the seed industry to change breeding objectives may also emerge as a constraint to progress in agricultural innovation and productivity. Before long, the focus of corn breeders may need to begin a steady shift from just increasing yields through more dense plantings to also preserving net per acre income by reducing costs more so than yields. Such a gradual evolution in breeding priorities will mean greater focus on using genetics to tighten the efficiency of key root system-soil-nutrient cycle-plant health linkages, and also has great potential in reducing environmental problems.

Consider the multiple consequences of a change in policy that in turn triggers the need for a basic shift in breeding priorities. Suppose that the nation's commitment to reduce carbon emissions leads to a willingness of government to support efforts to enhance terrestrial storage of carbon. Suppose a consensus emerges that changes in breeding objectives-for example, selecting for fuller, deeper root system architecture-are needed for corn production to cost-effectively compete with forestry in enhancing soil carbon levels.

Biotechnology has great potential to enhance the capacity of breeders to produce genetically improved varieties. It is legitimate and important to question whether a portion of potential societal benefits may be foregone because they are judged "not compatible" with steps seen as necessary to maximize future industry profits. This would be a classic case of market failure, where actions taken by a private company, acting in the best interest of its stockholders leave the nation less well off. Experience also shows that the bigger and more profitable an industry becomes, and the companies within it, the more difficult time government has in changing corporate behavior if and when conflicts arise between private profits and public needs.

This does not bode well for the future role of plant genetics, for years one of the public plus private sector's most reliable tools for problem-solving and productivity improvement. New efforts and priorities, and heightened investments in the private sector is just part of the problem. The other is the retreat of the public sector from a meaningful, sustained and competitive role in the development of agricultural technology.

CHAPTER 2

FOOD SAFETY ISSUES

Recombinant-DNA technology in modern plant breeding and biotechnological food production systems, efforts began to define internationally harmonized evaluation strategies for the safety of foods derived from genetically modified organisms (GMOs). Two years after the first successful transformation experiment in plants (tobacco) in 1988, the International Food Biotechnology Council (IFBC) published the first report on the issue of safety assessment of these new varieties. Such as the Organisation for Economic Cooperation and Development (OECD), the Food and Agriculture Organization of the United Nations (FAO) and the World Health Organization (WHO) and the International Life Sciences Institute (ILSI) have developed further guidelines for safety assessment which have obtained broad international consensus among experts on food safety evaluation.

ECONOMIC COOPERATION AND DEVELOPMENT ORGANISATION

In 1993 the OECD formulated the concept of substantial equivalence as a guiding tool for the assessment of genetically modified foods, which has been further elaborated in the following years (Figure 1). The concept of substantial equivalence is part of a safety evaluation framework based on the idea that existing foods can serve as a basis for comparing the properties of a genetically modified food with the appropriate counterpart. The existing food supply is considered to be safe, as experienced by a long history of use, although it is recognized that foods may contain many anti-nutrients and toxicants which, at certain levels of consumption, may induce deleterious effects in humans and animals.

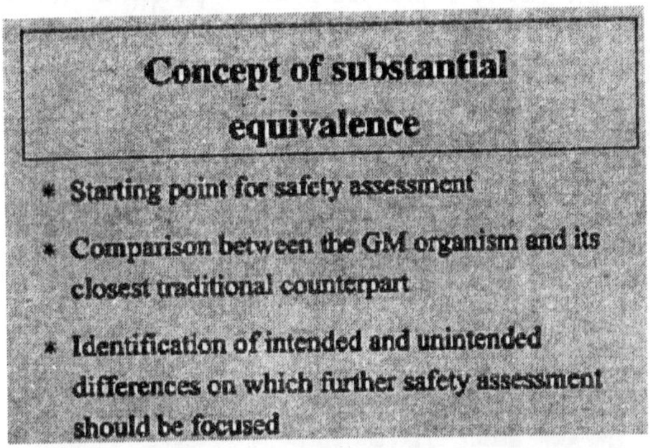

Figure 1: The concept of substantial equivalence.

Application of the concept is not a safety assessment per se, but helps to identify similarities and potential differences between the existing food and the new product, which is then subject to further toxicological investigation. Three scenarios are envisioned in which the genetically modified plant or food would be

— substantially equivalent;

— substantially equivalent except for the inserted trait; or

— not equivalent at all.

A compositional analysis of key components, including key nutrients and natural toxicants, is the basis of assessment of substantial equivalence, in addition to phenotypic and agronomic characteristics of the genetically modified plant.

In the first scenario, no further specific testing is required as the product has been characterized as substantially equivalent to a traditional counterpart whose consumption is considered to be safe, for example, starch from potato. In the second scenario, substantial equivalence would apply except for the inserted trait, and so the focus of the safety testing is on this trait, for example, an insecticidal protein of genetically modified tomato. Safety tests include specific toxicity testing according to the nature and function of the newly expressed protein; potential occurence of unintended effects; potential for gene transfer from genetically modified foods to human/ animal gut flora; the potential allergenicity of the newly inserted traits; and the role of the new food in the diet (Figure 2).

Safety issues of GM foods

(i) Genetic modification process
(ii) Safety of new proteins
(iii) Occurrence and implications of unintended effects
(iv) Gene transfer to gut microflora
(v) Allergenicity of new proteins
(vi) Role of the new food in the diet
(vii) Influence of food processing

Figure 2: Safety issues of genetically modifiied foods.

In the third scenario, the novel crop or food would be not substantially equivalent with a traditional counterpart, and a case-by-case assessment of the new food must be carried out according to the characteristics of the new product. Key components of a specific crop for comparison with a genetically modified crop are described by Consensus Documents compiled by the OECD's Task Force for the Safety of Novel Foods and Feeds. These documents provide useful guidance on which components should be minimally analysed.

INTERNATIONAL LIFE SCIENCES INSTITUTE

A consensus document has been prepared by ILSI Europe on

evaluation of the safety of novel foods. It provides background for data requirements for all novel foods, including foods and food ingredients derived from GMOs. For genetically modified foods this will include data on transgenic DNA; phenotype; and composition including gross composition, nutrients, anti-nutrients, and toxins. Substantial equivalence of the novel food to an appropriate counterpart can then be determined.

There is a degree of freedom in choosing the level at which this comparison should be carried out, such as the food source, food product, and molecular levels. Similar to the OECD's and FAO/WHO's consensus views, the ILSI document defines three scenarios in which the novel food or food ingredient is characterized as

— substantially equivalent to a reference food/ingredient;

— sufficiently similar; or

— not sufficiently similar.

For novel foods and novel food ingredients that are not substantially equivalent, nutritional and toxicological data, and data concerning allergenic potential, need to be considered. A decision tree for testing genetically modified foods for allergenicity has been developed by ILSI in collaboration with the IFBC. Three scenarios are considered where the source of the transgene may be:

— a commonly allergenic food;

— a less commonly allergenic food or other known food source; or

— without a history of allergenicity.

Criteria used in the decision tree include: source of the transferred material; sequence homology of the transgene product to allergenic proteins; immunoreactivity of the transgene product tested with sera from individuals who are allergic to the source; stability of the transgene product under gastro-intestinal conditions, or under heat or other processing conditions. IgE-binding tests are recommended with the new protein derived from known allergenic sources, using sera from individuals allergic to those sources, followed if necessary by skin-prick testing and double-blind food challenges.

The decision-tree approach for new proteins derived from sources with no known history of allergenicity relies primarily upon sequence homology comparisons with known allergens, and on the stability of the protein under gastro-intestinal and food-processing conditions.

FOOD AND AGRICULTURE ORGANIZATION/WORLD HEALTH ORGANIZATION

FAO and WHO have been organizing workshops and consultations on the safety of GMOs since 1990. At the Joint FAO/WHO Consultation in 1996 it was recommended that the safety evaluation should be based on the concept of substantial equivalence, which is 'a dynamic, analytical exercise in the assessment of the safety of a new food relative to an existing food'.

The following parameters should be considered to determine the substantial equivalence of a genetically modified plant: molecular characterization; phenotypic characteristics; key nutrients; toxicants; and allergens. The distinction between three levels of substantial equivalence of the novel food to its counterpart, and the subsequent decisions for further testing based upon substantial equivalence, are similar to those defined by OECD. The Codex Alimentarius Commission of FAO/WHO is committed to the international harmonization of food standards. Food standards developed by Codex Alimentarius should be adopted by the participating national governments. The Codex ad hoc Intergovernmental Task Force on Foods Derived from Biotechnology has the task to develop standards, guidelines and other recommendations for genetically modified foods.

Risk assessment covers issues such as food safety, substantial equivalence and long-term health effects, while risk analysis may include decision-making and post-market monitoring. For example; an Expert Consultation held in Geneva, Switzerland in May/June 2000 evaluated experiences gathered since the 1996 Consultation. Topics considered included substantial equivalence, unintended effects of genetic modification, food safety, nutritional effects, antibiotic resistance marker genes, and allergenicity. The Consultation endorsed the concept of substantial equivalence as a pragmatic approach for the safety assessment of genetically modified foods, and concluded that at present no suitable alternative strategies are available. Application of the concept is a starting point for safety assessment, rather than an endpoint. It identifies similarities and possible differences between the genetically modified food and its appropriate counterpart, which should then be assessed further.

The issue of the potential occurrence of unintended effects due to the genetic modification process, such as the loss of existing traits or the acquisition of new ones, was examined. The occurrence of unintended effects is not unique for the application of recDNA techniques, but also occurs frequently in conventional breeding. Present approaches to detecting such effects focus on chemical analysis of known nutrients and toxicants. In order to increase the possibility of detecting unintended effects, profiling/fingerprinting methods are considered useful alternatives. This is of particular interest for plants with extensive modifications of the genome where chances of the occurrence of unintended effects may increase. Animal studies are deemed necessary to obtain information on the characteristics of newly expressed proteins, analogous to the conventional toxicity testing of food additives.

Testing of whole foods may be considered if relevant changes in composition may have taken place in addition to the expected ones. The minimum requirement to demonstrate the safety of long-term consumption of a food is a subchronic 90-day study. Longer-term studies may be needed if the results of a 90-day study indicate adverse effects such as proliferative changes in tissues. The Expert Consultation noted that, in general, very little is known about the potential long-term effects of any foods, and that identification of such effects may be very difficult, if not impossible, due to the many confounding factors and the great genetic variability in food-related effects among the population.

Thus the identification of long-term effects specifically attributable to genetically modified foods is highly unlikely. Epidemiological studies are not likely to identify such effects given the high background of undesirable effects of conventional foods. The Consultation was of the opinion that pre-market safety assessment already gives an assurance that genetically modified foods are as safe as their conventional counterparts. Experimental studies, such as randomized controlled human trials, if properly performed, might provide additional evidence for human safety in the medium to long term. At the FAO/WHO Expert Consultation on Allergenicity held in Rome in January 2001, a new decision tree was developed. The new decision tree builds on the one developed by ILSI/IFBC and FAO/WHO.

When the source of the new protein is known to be allergenic, sequence similarity with known allergens and subsequent specific in vitro screening in sera from patients allergic to the source is recommended. Criteria for a positive outcome of sequence similarity include >35% identity in amino acid sequence between the expressed protein and a known allergen, or identity of six contiguous amino acids. In contrast to the previous decision trees, the new tree makes no distinction between commonly and less commonly allergenic sources with respect to in vitro screening. Any positive outcome defines the product as allergenic, and further product development should be discontinued. A negative outcome of the specific in vitro serum screening will lead to further targeted serum screening, testing the expressed protein for pepsin resistance and immunogenicity in animal models.

Targeted in vitro serum testing is done with sera from patients allergic to materials that are broadly related to the source of the original gene. Human in vivo testing may be considered in selected cases, but is not mandatory. In case the new protein comes from a source not known to be allergenic, the decision-tree approach focuses on sequence similarity with known allergens, and subsequent targeted in vitro serum screening to test for crossreactivity. Where there is sequence homology and the outcome of the serum-screening tests is positive, the protein is considered to be allergenic. Where the outcome is negative, further testing of the pepsin resistance of the new protein and immunogenicity testing in animal models may give indications for high or low probability of the allergenic potential of the new protein.

FOOD SAFETY REGULATIONS

The safety evaluation may focus on different levels of the food crop, for example, the whole crop; crop tissues; or purified products, depending on the scope of the applica-tion. In addition to foods and food ingredients, the use of purified products as food additives is also envisioned.

Food and food ingredients: European Union. In 1997 the Regulation on novel foods and novel food ingredients came into force in the European Union. This regulation distinguishes six categories of novel food products, two directly referring to products derived from GMOs. The concept of substantial equivalence is fully endorsed in the European approach. It is stated that the assessment of substantial equivalence is an analytical process, where the novel food is compared to the most appropriate approved food, not necessarily meaning a conventional food, but possibly an earlier approved genetically modified variety.

This analytical comparison to assess whether or not a novel food product is substantially equivalent to a product that is already on the market is, at the same time, the basis for both toxicological and nutritional assessments. If additional in vivo experiments are deemed necessary, it is stated to be essential to have sufficient knowledge on the nutritional characteristics of the novel food, for example, the energy content, protein content, and bioavailability of micronutrients.

The highest test dosage should be the maximum amount of novel food product that can be included in a balanced animal diet, while the lowest dosage should be comparable to the expected amount in the human diet. If desirable safety factors cannot be reached in this way, additional investigations on resorption and metabolism of the novel food in animals, and eventually humans, are required; however, in specific cases lower safety factors may be acceptable if additional data show the safety of the novel food. The exposure assessment should include specific vulnerable consumer groups.

For the nutritional assessment, it may be necessary in some cases to set up post-launch monitoring programmes. Also, with relation to allergenicity, the EU largely follows international consensus reports in that potential allergenicity should be investigated with the available means, to avoid the introduction of new allergens into the food supply. Thirteen decision trees are added to the regulation in order to guide producers to the data needed to establish the safety of an individual novel food product.

Outside the EU, foods from genetically modified crops fit into regulatory frameworks that differ from nation to nation. Under Canadian regulations, genetically modified crops are considered novel foods, similarly to the EU. Japanese and Australia/New Zealand's regulations, on the other hand, focus specifically on foods derived from genetically modified crops. The American regulations do not in principle regard genetically modified crops as a separate entity with respect to other foods.

Rather, the focus is on the altered characteristics brought about by genetic modification, and the intended use of the novel crop.

Food additives: Food additives derived from GMOs are regulated differently in Australia/New Zealand, Canada, the EU, the USA and Japan, and the definition of 'food additive' varies between these nations. In the EU and Canada, the evaluation of food additives – non-nutrient substances not conventionally present in food – does not distinguish between food additives derived from GMOs or from other sources. In Australia and New Zealand, food additives from GMOs are evaluated for the components that deviate from the existing specifications for food additives.

In the USA, a food additive is defined as a non-GRAS food component. Food components that are 'food additives' under EU legislation may therefore be considered either food ingredients (GRAS) or American 'food additives'. Commercial use of American food additives requires a permit following a safety evaluation by the Food and Drug Administration (FDA). Introduced gene products are considered food-additives, i.e. non-GRAS components, unless they have already been declared GRAS. In Japan, both genetically modified foods and food additives are subject to the same evaluation procedure.

CONCEPT OF SUBSTANTIAL EQUIVALENCE

The safety of our existing foods is based on long-term experience and history of safe use, even though they may contain anti-nutritional or toxic substances. In the past decades progress has been made with respect to identification and characterization of food constituents which may exert adverse and/or beneficial effects on chronic intake. Several compounds have been identified with anti-carcinogenic effects, and positive effects on osteoporosis and on the incidence of cardiovascular diseases, while certain plant compounds may also exert adverse mutagenic and carcinogenic effects.

In many cases beneficial compounds may also exert adverse effects, depending on the conditions and the presence of other agonists or antagonists. The scientific basis underpinning the relationship between food and its constituents and health is still fragmentary, but positive and negative effects due to the consumption of certain food constituents cannot be ignored. Thus the OECD concept of generally recognized safety of the existing food supply will undergo further refinement in the light of growing scientific evidence for the biological relevance of specific food constituents. It is generally acknowledged that the basis for the foodsafety evaluation of complex (plant) products should be a comparison with

i) the nearest comparator (in genetic terms); and

ii) other varieties that are already on the market, in that order.

This comparison comprises phenotypic characteristics and composition. The breeder usually performs the phenotypic comparison, but the criteria are not well defined. On the other hand, this type of comparison has been used successfully in plant breeding for decades, and has led to many new varieties with virtually no negative consequences for the consumer. The compositional analysis has been the subject of an ongoing discussion in many national and international meetings on the issue of the safety of GMO-derived products.

From the start of the practice of comparing constituents of genetically modified varieties with their traditional counterpart, it has been advocated that both nutrients and antinutrients should be included in the analysis. As mentioned above, the OECD has taken the lead in formulating Consensus Documents which group constituents that should be analysed, in all cases, in any new variety of the given crop. Additional analyses may be required, depending on the type of genetic modification or in order to further investigate detected differences. This should be determined on a case-by-case basis. International harmonization and standardization are necessary to avoid differences in data requirements in different countries and thereby to prevent trade barriers.

OECD Consensus Documents have now have been formulated for soybean and rapeseed, while others on maize, potato, sugar beet and rice are in the pipeline. The comparator for assessment of substantial equivalence will preferably be the direct parent line, which will not, however, be available in all cases. If the parent line is not available for comparison, the OECD advocates the use of several control lines to determine whether any observed differences may arise from secondary effects from the genetic alteration. Compositional analyses should be performed on genetically modified plants and their comparators that have been grown under similar environmental conditions, as these conditions may lead to differences in composition that are not related to the genetic modification.

At the same time, it is deemed necessary to assess the novel genetically modified variety at different locations and during subsequent growing seasons in order to assess whether other metabolic pathways may be turned on or switched off under different conditions, with possible effects on the composition of the food plant. The number of environmentally different locations where the genetically modified plant needs to be assessed is, in most cases, not specified. 'Standard' statistical analyses are usually performed on data for the genetically modified variety and the parent line, leading to acceptance or rejection of the hypothesis with a certain probability.

However, it is feasible that this system will need to be further elaborated in the future, as different aberrant compositional profiles may be acceptable for different groups of constituents. It can be envisioned that significant changes in metabolic pathways leading to toxic plant substances, such as glycoalkaloids in potato and tomato, will need to be investigated further – even if the changes do not lead to natural toxin levels that fall outside the natural variability ranges, as documented in the literature or determined from a traditionally bred group of control varieties.

On the other hand, it should be clear that in those cases where differences in composition between the modified organism and its counterpart fall outside these ranges of variability, such crops do not necessarily pose a threat to human health. Whether additional investigations are appropriate to address any further concerns related to the food safety of the crop plant should be assessed on a case-by-case basis. The Nordic Council proposed that, in the case of a difference of 20% in the average value for the new plant variety compared to the parent line, an explanation should be sought; but it is doubtful whether acceptable degrees of compositional differences can be defined in general.

RISK ASSESSMENTS

Safety assessment of food additives and food contaminants: The safety or risk evaluation of food additives, residues of pesticides and veterinary drugs, and food contaminants is based on

— hazard identification;

— hazard characterization; and

— assessment of exposure.

The potential of a chemical agent to cause harmful effect, and risk as a function of the probability that an adverse effect will occur due to the presence of a hazardous compound in food and the severity of the adverse effect. FAO/WHO and the International Program on Chemical Safety (IPCS) have developed strategies for the safety evaluation of these types of chemicals which may be present in food. These strategies focus on the establishment of a level of daily intake by humans, on a body weight basis, which would not cause an appreciable risk. The assumption is that for most toxic effects induced by chemicals, a threshold level can be determined, that is, a dose level below which a toxic effect is not apparent.

In order to arrive at such a dose, a battery of standardized toxicity tests is carried out: acute and (sub)-chronic toxicity studies, genotoxicity studies, carcinogenicity studies, and specific studies concerning immunotoxicity, reproduction and developmental toxicity. The protocols for such studies have been elaborated by OECD. From these studies, mostly carried out in laboratory animals, the NOAEL is determined and, upon application of a safety factor, the ADI is derived. In many cases a safety factor of 100 is used, allowing for differences in sensitivity between test animals and humans and for differences within the human population.

Depending on the available data and the substance under study higher or lower safety factors may be applied. The use of relatively large default factors in establishing an ADI probably provides an overestimation of the true risk involved, and can be considered as a 'safety first' approach. However, certain chemicals such as genotoxic carcinogens do not show a dose-dependent threshold level – in these cases the ADI concept is not applicable, and a quantitative risk assessment is carried out taking into account the incidence of DNA damage and tumours versus the applied dose.

As whole foods contain mixtures of macro- and micronutrients, anti-nutrients and plant toxins, the safety evaluation of foods for single, well defined chemicals is virtually impossible. Foods may contain toxic compounds, often with small margins of safety between actual intake and apparent toxic-effect levels.

For instance, the safety margin for potato glycoalkaloids may be between 2 and 6, assuming a lowest observed effect level in humans of 2 mg kg^{-1} body weight, an average level of glycoalkaloids in potato of 200 mg kg^{-1}, and an average daily consumption of 300 g. Moreover, for certain micronutrients margins may be small between levels that are beneficial or essential for human health, and levels that are toxic. For instance, the recommended dose of vitamin A is 1 mg day^{-1} for pregnant women, while the estimated safe daily intake level is 3 mg, and teratogenic effects have been observed at a daily intake of 7.5 mg day^{-1}.

In the case of essential amino acids, adverse effect levels in mg kg^{-1} body weight are only three to four times the nutritional requirements in humans. Application of the usual safety factor of 100 would result in inadequate nutritional levels. Testing of whole foods in laboratory animals has its specific problems, and considerable experience has been gained with toxicological testing of irradiated foods. The amounts of foods to be administered to animals are limited due to effects on satiety and possible negative interference with the nutritional balance of the animal diet. Feeding animals with whole foods at exaggerated dose levels may induce a series of adverse effects that would mask potential adverse effects caused by alterations induced by the genetic modification.

The highest test dosages should be the maximum amount of novel food product that can be included in a balanced animal diet, while the lowest dosage should be comparable to the expected amount in the human diet. Furthermore, the bio-availability of food constituents may be different when ingested as part of the food matrix. The minimum duration of an animal study with whole foods to demonstrate the safety of long-term consumption of a food depends on the available toxicity database for the food under investigation.

The 2000 FAO/WHO Expert Consultation recommended as minimum requirement a subchronic 90-day study, with possibly longer-term studies needed if the results of a 90-day study indicate adverse effects, such as proliferative changes in tissues. Further studies are needed to establish specific research protocols. Appropriate safety testing with whole foods should be hypothesis-driven and should be carried out in parallel with toxicity studies on specific, isolated food constituents. A combination of animal experiments, in vitro experiments with tissues and/or organs from animals and humans, and possibly human clinical studies, should be carried out, focusing on the identification of biomarkers for exposure and markers that are predictive in an early phase of exposure of chronic toxicity.

Role of Diet: For both traditionally bred and genetically modified varieties with a novel trait that either intentionally or unintentionally interferes with the nutritional characteristics of the crop, it will be very important to assess adequately the consequences of introducing this novel plant variety onto the market, for the nutritional status of specific consumer groups and the entire population. In order to do this effectively and identify possible consumer groups at risk of nutritional deficiencies, it will be necessary to have detailed information on the consumption patterns of different consumer groups within the population, and on geographic variations in these patterns.

Extensive databases are currently available only for a limited number of countries. The availability of such databases will become more compelling when more novel foods entering the market have significantly altered nutritional characteristics. New models to assess the exposure of individual consumers to individual foods and food ingredients will also gain importance. An example of such an approach is the Monte Carlo simulation approach. Monte Carlo models assess the distribution of exposure of individuals within a given population, taking into account the probability that exposures from more than one source may occur on a single day without overstating the actual exposure. This is especially relevant for both novel transgenic proteins that are introduced in different food crops and/or crop varieties, as well as for GMO-derived ingredients that may be used in a large variety of food products. To obtain reliable information from models of this type, it is necessary to collect sufficiently sound input parameters for the populations under investigation.

Safety evaluation of newly expressed proteins: If substantial equivalence can be established except for a single or few specific traits of the genetically modified plant, further assessment focuses on the newly introduced trait itself. Demonstration of the lack of amino acid sequence homology to known protein toxins/ allergens, and a rapid proteolytic degradation under simulated mammalian digestion conditions, was deemed to be sufficient to assume the safety of the new protein. However, there may be circumstances that require more extensive testing of the new protein, such as

— the specificity and biological function/mode of action of the protein is partly known or unknown;
— the protein is implicated in mammalian toxicity;
— human and animal exposure to the protein is not documented; or
— modification of the primary structure of naturally occurring forms. Bacterial Bt proteins are an example of proteins that have been introduced into crop varieties by genetic modification.

Bt proteins from Bacillus thuringiensis strains have been introduced into genetically modified crop plants for their insecticidal properties in the larvae of target herbivoral insect species. The working mechanism is based on specific receptor binding in susceptible insect larvae in epithelial cells of the midgut, leading to pore formation, cell lysis, disintegration of the epithelium lining in their midgut, and eventually to death of the larvae due to starvation. This type of biological action of the newly introduced protein directs further toxicity testing in mammals.

A general drawback is that newly expressed pesticidal proteins, such as Bt toxins and lectins, are often present in the genetically modified plant at levels too low for extensive testing. Therefore, sufficient amounts of the new proteins are obtained from cultures of overexpressing bacterial strains. This carries the potential hazard that toxic impurities can be present, and that protein processing, like glycosylation, may be different in plants and bacteria. Therefore it is important to demonstrate that production in an alternative host does not result in differences in toxicity. For these pesticidal proteins, the following properties must be comparatively investigated:

— electrophoretic behaviour of full-length as well as trypsinated forms;
— immunoreactivity with poly- and/or monoclonal antibodies;
— identical patterns of posttranslation modification;
— sequence similarity; and
— functional characteristics to target insect species.

The safety of a number of newly inserted proteins has been tested on a case-by-case basis. It should be noted that for transgenic viral proteins in crops approved in Canada and the USA, their consumption has been assumed to be safe based on the history of ingestion of the wild-type plant viruses contained within plant foods.

In the case of the Cry1Ab5 and Cry9C proteins, various studies have been performed on binding to tissues of the gastro-intestinal tract of rodents and primates, including humans. There is no evidence for the presence of specific receptors in mammalian tissues for these proteins, nor are there indications of an amino acid sequence homology to known protein toxins/food allergens. A number of toxicity tests have been performed with respect to: digestibility and stability in in vitro simulated gastric and intestinal fluids and in vivo models; acute oral toxicity in a rodent species; subchronic toxicity with focus on a tier I immunotoxicity screening.

Experiments performed with single and repeated dosing of the Cry proteins Cry1Ab5 and Cry9C, at levels up to 10 000 times those produced in genetically modified plants, did not indicate toxic effects in the rat, and histopathological analysis did not show binding of the Cry proteins to the intestinal epithelium of rodents and tissues of other mammals. In contrast to Cry1Ab5, Cry9C showed resistance to proteolysis under simulated human gastric conditions ($pH > 2.0$) and denaturation at elevated temperatures. On the other hand, it was noted that Cry9C degraded completely upon pepsin treatment at pH <1.5. However, the digestibility of protein preparations under simulated conditions is of limited value, as questions can be raised as to whether these assays do mimic the physiological state of human beings. In cases of

— a completely novel gene;

— novel proteins as anti-nutrients;

— novel proteins without a clear threshold;

— predicted high levels of intake of toxic proteins such as protease inhibitors; and

— non-rapidly degradable proteins,

more extensive toxicity testing with the pure protein at exaggerated doses may be required.

Allergenicity: The potential allergenicity of newly introduced proteins in genetically modified foods is a major safety concern. This is true in particular for genetic material obtained from sources with an unknown allergenic history, such as the soil bacterium B. thuringiensis. An illustrative case of a genetically modified food for which the allergenic risk has to be assessed is maize in which the truncated Cry9C protein is expressed, and which has been allowed as a transgene product in StarLink yellow maize for animal feed in the USA. It should be noted that the protoxin Cry9C from B. thuringiensis var. tolworthi has been modified at residue 164 by substituting the arginine residue with lysine to increase serine protease resistance in the field. Recent investigations have found traces of the Cry9C gene and/or protein in taco shells.

The Cry9C protein has also been detected in maize seeds of a non-StarLink variety or in maize from such seeds. This has raised the issue of potential allergenicity of the genetically modified maize for humans. Cry9C might be a potential allergen because the protein shows some characteristics of known food allergens:

— an MW of 68 kDa;

— relative resistance to gastric proteolytic degradation and to heat and acid treatment;

— it is probably a glycoprotein;

— induces a positive IgE response in the Brown Norway rat, and is a high IgE responder on intraperitoneal and oral sensitization, in contrast to the related Cry1Ab5 protein; and

— may be found intact in the bloodstream after oral feeding in a rat model.

Cry9C has no amino acid sequence homology to any known allergen or protein toxin, and wild-type and StarLink maize protein extracts have been demonstrated to be indistinguishable in their reactivities towards sera of maize-allergic and major food-allergic patients. Furthermore, no immunogenic/toxic effects were observed in a 30-day repeated-dose study in mice with Cry9C, and the bioavailability of the protein in the rat is relatively low, which reduces the likelihood of sensitization.

Levels of Cry9C in maize-derived food products appear to be much less than the >1% level apparently characteristic of food allergens. Post-harvest blending and mixing may have diluted the Cry9C protein in food products to the p.p.b. range for the harvest years 2000 and 1999. Maize in food channels is either wet-mill processed, which produces high-fructose corn syrup, glycose, dextrose, starch or oil; or dry-milled, which produces primarily cereals, flour and meal. A preliminary study using Cry9C ELISA well tests showed that there was no intact Cry9C protein in a limited number of starch samples. In general, the protein fraction goes to feed use. Upon drymilling, the Cry9C protein content is reduced by 40%.

Additional processing, such as alkaline cooking, decreases the protein content to 0.1–0.2% of the original Cry9C protein. This suggests a further reduction in allergenic potency; however, protein denaturation by heat or partial proteolysis may uncover new allergenic epitopes. It is therefore important to note the need for reproducible, validated methods for analysing Cry9C protein levels in processed foods and intermediates, as distinct from the PCR methods. The estimated duration of exposure to Cry9C is uncertain, but may have been too short to promote sensitization and induce allergenic reactions.

An example of a transgene from an allergenic source is that of the Brazil nut 2S albumin expressed in soybean. This protein is rich in methionine, and would therefore increase the nutritive value of soybean for animal feed. It was found, however, that the transgenic protein was reactive towards sera from patients who were allergic to Brazil nut, and the further development of this soybean was halted.

Detection and characterization of unintended effects: Upon random insertion of specific DNA sequences into the plant genome, the disruption, modification or silencing of active genes or the activation of silent genes may occur, which may result in the formation of either new metabolites or altered levels of existing metabolites.

Unintended effects may be partly predictable on the basis of knowledge of the place of the transgenic DNA insertion, the function of the inserted trait, or its involvement in metabolic pathways; while other effects are unpredictable due to the limited knowledge of gene regulation and gene–gene interactions. It should be emphasized that the occurrence of unintended effects is not specific for genome modification through recDNA technology—it also occurs frequently in conventional breeding. Unintended effects may be identified by an analysis of the agronomical/morphological characteristics of the new plant and an extensive chemical analysis of key nutrients, anti-nutrients and toxicants typical for the plant.

Limitations of this analytical, comparative approach are the possible occurrence of unknown toxicants and anti-nutrients, in particular in food plant species with no history of (safe) use;

and the availability of adequate detection methods. Examples of unexpected secondary effects due to either somaclonal variations, pleiotropic effects or genetic modification, which may be of biological or agronomic importance to the plant. In order to identify potential secondary effects of the genetic modification, which would result in alterations in the composition of genetically modified crops, different strategies may be applied, for example the targeted approach, or the non-targeted approach.

Targeted approach using single compound analysis: For any given transformation event, targeted studies should include baseline analyses of a number of key nutrients such as proteins, carbohydrates, fats, vitamins and other nutritional/anti-nutritional compounds which, if unintentionally modified, might affect nutritional value and safety. Selection of key nutrients and toxicants needs to take into account the target species, structure and function of the inserted gene(s), and possible interferences in metabolic pathways (Figure 3). Selection of compounds may be limited to a restricted number which represents essential biochemical/physiological pathways in the organism. It is plausible, but not proven, that expected changes in the metabolism as a possible result of the genetic modification will be identified by analysis of a great number of components, but unexpected changes are merely identified by chance.

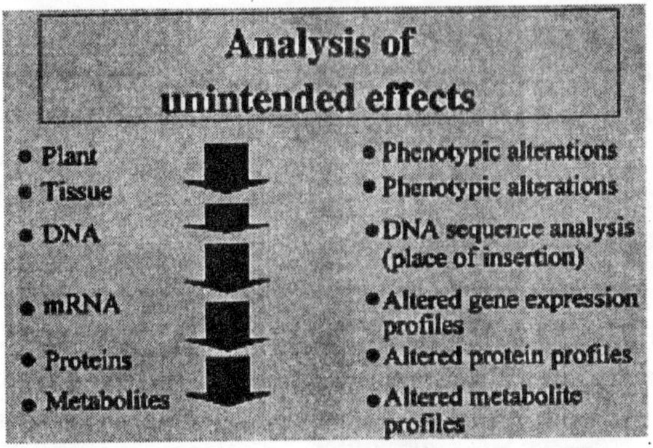

Figure 3: Different integration levels for the detection of unintended effects.

The targeted approach has severe limitations with respect to unknown anti-nutrients and natural toxins, especially in less well known crops.

Non-targeted approach using profiling methods: An alternative approach for the detection of unintended effects is the use of so-called profiling techniques. New methods are being developed which allow for the screening of potential changes in the physiology of the modified host organism at different cellular integration levels: at the genome level; during gene expression and protein translation; and at the level of metabolic pathways. Many factors, such as genetic characteristics; agronomic factors; environmental influences; plant—microbe interactions; maturity stage; and post-harvest effects determine the morphological, agronomic and physiological properties of a food crop. Screening for potential changes in these characteristics in genetically modified plants becomes more important as the newer genetic alterations changing agronomical or nutritionrelated properties are more complex, involving insertion of large DNA fragments or clusters of genes.

DNA analysis: Localization and characterization of the place of insertion are the most direct approaches to predicting and identifying possible occurrence of (un-) intended effects due to transgene insertion in recipientplant DNA. Data for transgene flanking regions will give leads for further analysis, in the case of a transgene insertion within or in the proximity of an endogenous gene. Transgene chromosomal location and structure can be detected by various methods such as genomic in situ hybridization and fluorescence in situ hybridization, and by direct sequencing of flanking DNA.

Knowledge of plant genomes is still limited, including the reliability of annotations in genomic databases, but the understanding of the genomic code and the regulation of gene expression in relation to the networks of metabolic activity is increasing. Therefore, the sequencing of the place of insertions will become increasingly informative.

Gene expression analysis: The DNA microarray technology is a powerful tool to study gene expression. The study of gene expression using microarray technology is based on hybridization of mRNA to a high-density array of immobilized target sequences, each corresponding to a specific gene. mRNAs from samples to be analysed are labelled by incorporation of a fluorescent dye and subsequently hybridized to the array. The fluorescence at each spot on the array is a quantitative measure corresponding to the expression level of the particular gene. The major advantage of the DNA microarray technology over conventional gene profiling techniques is that it allows smallscale analysis of expression of a large number of genes at the same time, in a sensitive and quantitative manner. Furthermore, it allows comparison of gene-expression profiles under different conditions.

The technology and the related field of bioinformatics are still in development, and further improvements can be anticipated. The potential value of the application of technology for the safety assessment of genetically modified food plants is currently under investigation. The tomato is used as a model crop. To study differences in gene expression, two informative tomato expressed-sequence-tag (EST) libraries are obtained, one consisting of ESTs that are specific for the red stage of ripening, and the other for the green, unripe stage. Both EST libraries are spotted on the array and, in addition, selected functionally identified cDNAs, selected on the basis of their published sequence.

The array is subsequently hybridized with mRNAs that are isolated from a number of different genetically modified varieties under investigation, as well as with the parent line and control lines. Preliminary results show that reproducible fluorescence patterns may reveal altered gene expression outside the ranges of natural variation, due to different stages of ripening. Prospects are that this method may effectively be used to screen for altered gene expression and, at the same time, provide initial information on the nature of detected alterations, whether the observed alteration may affect the safety or nutritional value of the food crop under investigation.

Proteomics: Correlation between mRNA expression and protein levels is generally poor, as rates of degradation of individual mRNAs and proteins differ. Therefore, understanding of the biological complexities in the plant cell can be expanded by exploiting proteomics, a technique that analyses many proteins simultaneously and will contribute to our understanding of gene function. Particularly, recent developments in mass spectrometry have increased the applicability of two-dimensional gel electrophoresis in the studies of complex protein mixtures. Proteomics can be divided into three main areas:

i) identification of proteins and their post-translational modifications;

ii) 'differential display' proteomics for quantification of the variation in contents; and

iii) studies of protein–protein interactions.

The method most often used for analysing differences in protein pattern is sodium dodecyl sulfate–polyacrylamide gel electrophoresis, followed by excision of protein spots from the gel, digestion into fragments by specific proteases, and subsequently analysis by mass spectrometry. It allows the identification of proteins by comparing the mass of peptide fragments with data predicted by genetic or protein sequence information. Other much faster technologies, such as protein chip-based approaches, are under development.

In addition, major technical hurdles remain to be overcome: proteins may constantly change in their secondary, tertiary and quaternary structures, depending on transfer and expression in different tissues and cellular compartments, which may profoundly influence their electrophoretic behaviour and molecular mass. When searching for unintended changes by 2-D PAGE, the first step is to compare proteomes of the lines under investigation. If differences in protein profiles are detected, normal variations should be evaluated. If the profiles are outside normal variations, identification of the protein must be carried out, which may lead to further toxicological studies.

Moreover, metabolic changes may be looked at if the identified protein has a known enzymatic activity. There is one example of the use of proteomics to study alterations in the composition of a genetically modified plant, which illustrates that a targeted change in the level of a specific protein can result in other proteins being affected. The improvement in rice storage proteins by antisense technology resulting in low-glutelin genetically modified rice for commercial brewing of sake has been associated with an unintended increase in the levels of prolamins. This would not have been detected by standard analyses such as total protein and amino acid profiling, but was observed only following SDS–PAGE. Prominently resolved polypeptide bands showed sequence homology with a number of known anti-nutrient and inhibitory proteins, which may have implications for the safe use of these seeds as human food.

Chemical fingerprinting: A multi-compositional analysis of biologically active compounds in plants–nutrients, antinutritional factors, toxicants and other relevant compounds—may indicate whether intended and/or unintended effects have taken place as a result of genetic modification. The three most important techniques that have emerged are gas chromatography (GC), high-performance liquid chromatography (HPLC) and nuclear magnetic resonance (NMR). These methods are capable of detecting, resolving and quantifying a wide range of compounds in a single sample.

For instance, metabolic profiling of isoprenoids by an HPLC method was described recently with applications to genetically modified tomato and Arabidopsis. The potential of GC as a metabolic profiling method for plants was demonstrated some 10 years ago, and GC/MS has been established as the most versatile and sensitive profiling method in the past 2 years following its systematic development. Determination of a chemical fingerprint was based on the detection of alterations in 1H-NMR spectra obtained from different water and organic solvent extracts from genetically modified tomato varieties, such as the antisense RNA exogalactanase fruit, and from their non-modified counterpart,

Application of these techniques will provide more detailed information on possible changes than can be obtained from single-compound analysis. Once differences have been identified, further safety evaluation of the observed differences may be needed by specific in vitro and/or in vivo testing. The design of such experiments will focus on the differences observed with the profiling methods. However, a number of problems must be addressed before such methods can be used on a routine basis:

— standardization of sample collection, preparation and extraction procedures;
— standardization and validation of measurements;
— limited availability of data on profiles and natural variations; and
— lack of bioinformatic systems to treat large data sets.

Currently, different methods are tested for the detection and characterization of unintended effects as a result of genetic modification.

Marker Genes: Code for resistance to herbicides or antibiotics are the most commonly used marker genes. The use of herbicide-resistant genes can be twofold: for selection purposes; and/or for altering the agronomic characteristics of a plant. In particular, the use of antibiotic-resistance genes is subject to controversy and intense debate, because of the risk of transfer and expression in bacteria which could compromise the clinical or veterinary use of certain antibiotics. Risk assessment of selectable marker genes with respect to the consumption by humans and animals of genetically modified foods or feed should focus, as with any new gene transfer, on micro-organisms residing in the gastro-intestinal tract of humans and animals, on the toxicity and allergenicity of newly expressed proteins, and on the impact of horizontal gene transfer.

Health aspects of marker genes have been dealt with by, among others, WHO; the Nordic Council; FAO/WHO(1996); FAO/WHO (2000b). There is general agreement that transfer of antibiotic resistance genes from plants to micro-organisms residing in the human gastro-intestinal tract is unlikely to occur, given the complexity of steps required for gene transfer, expression, and impact on antibiotic efficacy (FAO/WHO, 1996). Under conditions of selective pressure, a selectable marker may provide selective advantage to the recipient micro-organism. Transfer of plant DNA to microbial or mammalian cells would require the following steps:

— release of specific genes in the plant DNA;

— survival of the genes under gastro-intestinal conditions;

— competitive uptake of the genes;

— recipient bacteria or mammalian cells must be competent for transformation, and genes must survive restriction enzymes;

— insertion of the genes into the host DNA by rare repair or recombination events.

There are no data available indicating that marker genes in genetically modified plants transfer to microbial or mammalian cells. Transfer and expression of plant genes in bacteria have been observed under laboratory conditions, and only when homologous recombination was possible. This would imply that an antibiotic resistance-marker gene is introduced from plants into bacteria only if the same gene or other genes with identical sequences were present in the bacteria.

Model experiments with mice indicated the transfer of bacterially derived DNA fragments into mouse cells. These results have been criticized, along with others, regarding possible artefacts created during the analysis of foreign insertions in leukocyte DNA. A relevant consideration for the assessment of horizontal gene transfer, if it occurs, is the consequences of the transfer. Information must be available on the role of the antibiotic in human and veterinary use, its specific therapeutic spectrum, existing resistance levels in the environment, and possible alternatives for treatment of diseases.

The 2000 FAO/WHO Consultation concluded: 'For certain antibiotic resistance genes currently in use in genetically modified plants, available data suggest that consequences of horizontal gene transfer will be unlikely to pose a significant threat to the current therapeutic use of the respective drugs. With other genes that confer resistance to drugs that are important in specific medical use, or to drugs that have limited alternative therapies, the possibility of transfer and expression of these genes is a risk that warrants their avoidance in the genomes of widely disseminated GMOs and foods and food ingredients'. It then goes on: 'In future developments, the Consultation encourages the use of alternative transformation technologies, if available and demonstrated to be safe, that do not result in antibiotic resistance genes in genetically modified foods.

If further development of alternative technologies is required, additional research should be strongly encouraged'. Non-antibiotic marker genes such as tryptophane decarboxylase, b-glucuronidase and xylulose/ phosphomannose isomerase should be evaluated according to the characteristics of the newly encoded proteins and metabolites formed as result of enzymatic reactions. Furthermore, the risks of the presence of multiple markers and of multiple copies of markers should be evaluated. In one example, the isopentenyl transferase gene for plant hormone production allows modified cells to form shoots when cultured in dexamethasoneenriched media after the modification event.

Another way is to use the xylA gene, which encodes xylose isomerase, enabling the genetically modified plant cell to grow in cultures with the sugar xylose added, which is normally toxic to the plant cells. In addition, methods have been developed to excise genes after successful introduction, such as the CreLox system in which Cre is an enzyme that removes the stretch of DNA flanked by the Lox sequences. After successful transformation, expression of the recombinase gene was induced, and the marker and recombinase genes were subsequently removed by the recombinase. New models for safety testing, detection of unintended effects, gene transfer, detection and traceability of genetically modified foods are currently under development in the EU-funded research and technology development.

POST MARKETING SURVEILLANCE

Current practice in the pharmaceutical sector cannot be used as a model for PMS in the food sector, as the physician or other medical professional usually plays a crucial role in the collection of data on adverse effects of new pharmaceutical products. In addition, pharmaceutical products are separately packaged, usually taken by the patient during a limited time-frame, and patients will in many cases already be prepared for some adverse side-effects of the medication. These factors will enable adverse effects to be linked more easily to ingested medicines than to food products or ingredients.

Different strategies for post-marketing surveillance in the food-producing sector are available for food products that can easily be traced and identified. These methods vary from direct consumer feedback to the repurchase of products to determine the quality of the product on the shelf. These PMS strategies for food products will, in most cases, not be directly applicable to GMO-derived food products. Most products derived from a genetically modified plant will be used in products with slight changes in the recipes, depending on the genetically modified plants that are available to the producer. As these changes will, in most cases, not be reflected on the label, it will be difficult for the consumer to relate adverse effects to the specific ingredient or GMO component of an ingredient.

Only in the case of a genetically modified plant with an added value that the producer would like to communicate to the consumer, an identity-preserved food-production chain with constant composition and clear labelling may enable the consumer to trace any adverse effects back to the product. Adequate GMO detection and identification methods, in combination with repurchasing strategies, may supply comparable information on complex food products possibly containing specific genetically modified varieties, but this approach will be too elaborate and costly for the routine application that is necessary to be meaningful.

Only acute effects that are associated with high intakes of a substance are likely to become visible by PMS as, in general, long-term and/or rare effects usually require targeted epidemiological techniques beyond any normal post-marketing data collection. This will be even more valid for GMO-derived products outside identity-preserved food-production chains. Some cases of PMS in relation to food constituents have already been documented. Examples include the food additive aspartame, a high-intensity sweetener; and Olestra, a fat substitute used in snack foods. In the case of aspartame, the primary goal was to obtain more reliable information on the actual intake of the additive in comparison to pre-marketing projections.

Questions in relation to (unpredictable) allergenicity and alterations in the nutritional status of consumers as a result of the marketing of a particular novel food may be answered by PMS. A major challenge, however, will be to set up informative PMS systems for products that have not been monitored or surveyed so far, which will be relatively difficult to trace, and will participate in different foodproduction chains on a variable basis.

A number of genetically modified plants and foods obtained through extensive genetic modification(s) with the purpose of improving agronomic or food-quality traits will soon enter the commercial market. With respect to safety assessment, these new products should, in principle, also be assessed applying the concept of substantial equivalence. The recipient species in many cases provides a relevant baseline for the safety evaluation. For instance 'golden rice', with the b-carotene biosynthesis pathway introduced into the endosperm, contains genes from Narcissus pseudonarcissus coding for phytoene synthase and lycopene-cyclase under control of the rice glutelin promotor, as well as a bacterial gene from Erwinia coding for phytoene-desaturase under the 35S promotor.

β-Carotene was predominantly present, followed by the unexpected presence of the xanthophylls lutein and zeaxanthin. Non-modified isogenic rice functions as a comparator to identify potential changes in the composition, which should then be further assessed. The extent and target of the genetic modification, and the resulting alterations in metabolic pathways in the modified organism, guide the safety assessment and may lead to a more extensive toxicological safety evaluation compared to the genetically modified products that are now commercially available (OECD, 2001b).

The assessment of genetically modified plants/foods with enhanced nutritional properties should focus on the simultaneous characterization of inherent toxicological risks and nutritional benefits. This requires an integrated, multidisciplinary approach, incorporating molecular biology, toxicology, nutrition and genetics (Figure 4).

Issues to be addressed are:

— evidence for nutritional/health claims and target populations;
— toxicological and beneficial dose ranges of selected compounds;
— impact on overall dietary intake and associated effects on consumers;

— interactions between food constituents and food matrix effects; and
— possibilities for effective post-market surveillance, if necessary.

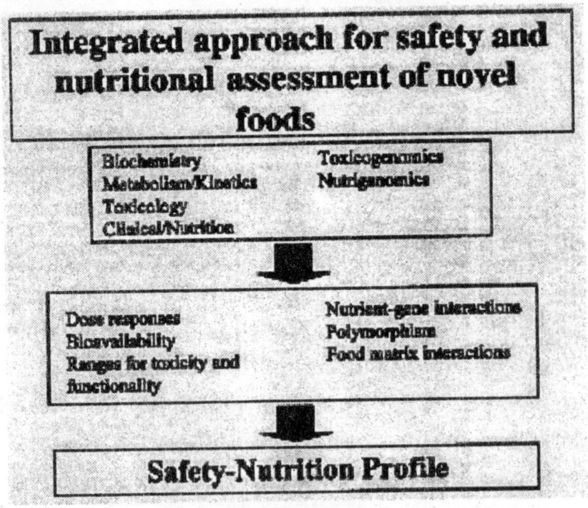

Figure 4. Integrated approach for safety evaluation of genetically modified foods.

Assessment of the safety of this type of foods is the crucial part of the evaluation, regardless of the potential benign effects of certain food constituents.

New innovative techniques such as the DNA microarray technology and proteomics are needed in order to characterize the complex interactions of bioactive food components at the molecular and cellular levels. Large-scale screening of the simultaneous expression of a large number of genes and synthesized proteins will provide relevant information concerning the complex relationships between human/animal exposure to bioactive food constituents and their specific effects. Moreover, insight can be gained in individual variabilities in biological responses, as well as in food matrix oriented interactions.

Whenever changes are made in the way of food production or processing, or when new foods without a history of use enter the market, a full safety and nutritional assessment with respect to implications for the consumers should be made. Various regulations have defined categories of foods and new food-processing methods which require such a safety assessment. The safety assessment of conventional crops is primarily based on analysis of agronomic performance and a by definition-limited analysis of known macro- and micronutrients, anti-nutrients and toxicants. Products with an unusual agronomic performance, taste, or harmful levels of specific compounds are rejected from the traditional breeding programme, for example, potato with high glycoalkaloid content, squash and zucchini containing cucurbitacin E, and celery containing furanocoumarins.

A long history of traditional breeding has given insight into the presence of nutritionally beneficial com-pounds and of anti-nutrients and toxicants in food plants, in which levels have been increased and/or diminished, respectively, through extensive breeding. This approach has great value and has resulted in a healthy and relatively safe food package, and should still be the leading principle when assessing the safety and wholesomeness of traditionally bred food crops. In the case of new plant varieties developed with traditional techniques with no appropriate comparator or history of safe use, application of the new profiling techniques is of great value for the assessment of the safety of these crops.

Chapter 3

Biotechnology and Foods

The terms "rDNA biotechnology" and "rDNA biotechnology-derived foods" to describe the application of recombinant DNA, or rDNA, technology to the genetic alteration of plants and microorganisms, and foods made therefrom. This technology, commonly known as genetic modification or gene splicing, allows for the effective and efficient transfer of genetic material from one organism to another.

Instead of cross-breeding plants for many generations or introducing mutations to introduce a desired trait-processes that are imprecise and that sometimes introduce unwanted changes-scientists can identify and insert one or more genes responsible for a particular trait into a plant or microorganism with greater precision and speed, although the current technology produces gene insertions at random locations. These transferred genes, or transgenes, do not have to come from a related species in order to be functional, and can be moved virtually at will among different living organisms.

Biotechnology

Biotechnology in the broad sense is, in fact, not a discrete technology. It refers to a group of useful enabling techniques, including but not limited to genetic modification, that have wide application in research and commerce. Over the past several decades, such techniques have become so integrated into the practice of plant breeding and microbiology and so commingled with conventional techniques as to blur distinctions between "old" and "new."

A useful working definition of biotechnology used by several United States government agencies is the application of biological systems and organisms to the production of useful goods and services. These encompass advances in biology, genetics, and biochemistry to technical and industrial processes as different as drug development, fish farming, forestry, crop development, fermentation, and oil spill clean-up. Turning to food biotechnology, the history of the development of modern genetics and molecular biology, which underpins much of this technology.

This pivotal demonstration was the first description of transformation, a mechanism of gene transfer that involves the uptake and integration of isolated DNA by an organism. It is a phenomenon that is central to an understanding of rDNA biotechnology, and may even be said to mark the beginning of the concept of the new biotechnology. Geneticists had earlier recognized that the chromosomes, linear structures composed of DNA and protein, were the vehicles of inheritance in the sense that they carried genes determining inherited characteristics.

Genes were conceived of as beads on a string. Genes that encode similar functions in different organisms are called orthologs, and genes that have the same structure in different organisms are said to have synteny. Many organisms are diploid, that is, they have two sets of chromosomes, one inherited from each parent. The pairs of chromosomes are present, in a constant and characteristic number, in all the cells of an organism. When the cells divide, the chromosomes also divide equally, by a process called mitosis.

When a diploid organism prepares for sexual reproduction by forming gametes, a reduction division, called meiosis, reduces the number of chromosomes so that each egg or sperm cell has exactly half the diploid number. At meiosis, there is a random assortment of maternally and paternally derived chromosomes, which is further complicated by exchanges between paired homologous chromosomes due to "crossing over" that takes place between chromosomes. Thus, in a sense, the genetic constitution of each gamete resembles a hand of cards dealt from a well-shuffled deck. In nature, gametes (germ cells) generally unite randomly at fertilization to restore the diploid condition.

Plant breeders use this variation by selecting the best plants that result from these combinations and stabilizing them by inbreeding or propagating them vegetatively. Thus, sexual reproduction produces "recombinant" organisms, in the sense that the organisms possess DNA rearranged and combined from two separate organisms. The task of plant and animal breeders is to select individuals that retain in a heritable way the desirable features of the parent lines. The segregation of genes with easily detected effects, such as round versus wrinkled seeds, was observed by Mendel, who first described the discrete nature of inheritance in peas.

Twentieth-century plant breeding, even before the advent of modern rDNA biotechnology methods, sought ways to take advantage of useful genes and gradually has found a wider and wider range of plant species and genera on which to draw. Breeders have long used interspecies hybridization, transferring genes between different, but related, species. Subsequently, plant geneticists found ways to perform even wider crosses between members of different genera using tissue culture techniques.

They include familiar and widely used varieties of tomato, potato, corn, oat, sugar beet, bread and durum wheat, rice, and pumpkin. Although DNA was known to play a key role in inheritance, it was not until Watson and Crick described the structure of the double-stranded DNA molecule that scientists understood how the exact replication of the DNA occurred at each cell division and how the sequence of nucleotides in the DNA molecule determined the sequence of nucleotides in messenger ribonucleic acid (mRNA) and in turn, through a triplet code, the sequence of amino acids in a protein.

When the DNA sequence of a gene is expressed, it is transcribed to form a single- stranded mRNA molecule, which is translated to make a protein. It is now known that the instructions for programming programming the development of a fertilized egg cell, or zygote, into an adult organism composed of millions of cells carrying identical sets of genes are encoded in the nucleotide sequence of the DNA. This is in the form of a code based on the four nucleotides, adenine, thymine, cytosine, and guanine, which form a series of three-letter words, or codons, that specify the amino acid sequences of the many thousands of proteins that carry out the cellular functions.

Biochemists have established that the basic metabolic events in all organisms have far more in common than was previously suspected. They found that not only is DNA the universal code used by all living things, but that the central functions of all organisms are nearly identical. DNA and ribonucleic acid (RNA) replication, protein synthesis, photosynthesis, energy metabolism, and a host of other functions were found to have much in common throughout living systems.

Molecular biologists soon learned to determine the sequences of genes that encoded these properties. As more and more genes were sequenced and compared, scientists found that the products of the genes that encode similar traits in very diverse organisms are often very similar in protein sequence. It also became apparent that most genes do not have characteristics specific to the organism in which they are found. In fact, it is impossible to determine the organism from which a gene arises by inspection of the gene sequence alone, although codon usage does vary among major groups of organisms.

Put another way, there is no way to identify "fish genes," "tomato genes," or "broccoli genes." The uniqueness of organisms instead lies not only in the DNA sequences of their genes, but also the organization of the genes which are present, and at what time and to what extent they are expressed. Enormous quantities of DNA have now been sequenced for a wide range of organisms. The genomes of several bacteria and small organisms have already been fully sequenced, and the genome sequences of higher organisms such as plants, insects, animals, and humans will soon be available.

In fact, about 40 genomes are expected to have been sequenced by the end of 2000. Even sequencing of the human genome is now more than 90% complete. One key observation is that, in the course of determining DNA sequences, identical genes are regularly found in organisms that are only remotely related. This observation has provided evidence that genetic transfer has occurred in nature to produce natural rDNA-containing organisms.

A discovery important to modern rDNA biotechnology techniques was the recognition that a series of so-called "restriction enzymes," thought to protect cells from invading viral DNA, could be used to cut the DNA at precise sites defined by the sequence of four, five, or six nucleotides at the site where the cut would be made. By using DNA ligases-enzymes that fuse together two pieces of DNA-the pieces of DNA formed by cutting DNA with restriction enzymes could be joined together into a single piece of DNA.

The fragments or pieces of DNA could also come from two different organisms. Pieces of DNA from different organisms are often called "heterologous DNA" and when heterologous fragments of DNA are joined together by a ligating enzyme, the fragment of DNA is said to be a "recombinant" molecule; i.e., it recombines DNA from two heterologous sources. The word "recombinant" is used analogously to describe the recombination of DNA of the parental chromosomes that takes place during meiotic cell division.

This ability to splice together pieces of heterologous DNA means that it is possible to clone fragments of DNA by splicing them into a bacterial plasmid, a circular self-replicating DNA molecule that multiplies inside the bacterial cell when it is introduced into the bacteria by a process called transformation. If the heterologous DNA was spliced into a site on the plasmid where the DNA would have an opportunity to be transcribed to mRNA, and then translated to form a functional and active protein, its action in the cell can be detected so that the function of the cloned fragment can be identified.

By this means, it is possible to produce very large numbers of copies of a known DNA fragment that can then be used to transform other organisms, such as plants and animals. Two methods of plant transformation are in use at the present time. One method, known as the ballistic or free DNA method, uses a gun to shoot microscopic particles of gold or tungsten into cultured plant cells. The particles are first coated with the DNA carrying the gene of interest, isolated from the bacteria in which it has been cloned.

Then, these particles are accelerated by releasing a charge of helium under high pressure. A small proportion of the particles penetrate not only the plant cell wall but the nuclear membrane as well. The DNA carried by these particles can be taken up and integrated into plant chromosomes. Although the entire nucleotide sequence of the segment of DNA to be introduced is usually known with the free DNA method, the site where the DNA is integrated cannot be predicted. While the sequence of the starting DNA can be determined with precision, free DNA delivery frequently leads to integration of multiple copies or portions of the gene of interest.

Selectable markers, i.e., genes whose expression can be detected soon after the cells have been treated with DNA, are used to recover the very small fraction of cells that are transformed. For example, if the markers confer resistance to a toxic agent, such as an antibiotic or a herbicide added to the culture medium, then only those cells which carry and express the non-host DNA are able to grow. Employs the bacterial plant pathogen Agrobacterium tumefaciens.

In nature, this bacterium infects wounds in broad-leafed plants and induces the formation of tumors or galls. The mechanism of tumor induction using the Agrobacterium method involves the movement of a part of the DNA of a large plasmid carried by the bacterium into some of the host cells. In some of the cells, a host cell chromosome takes up a part of the plasmid DNA, whereupon the plasmid DNA directs the cell to undergo repeated divisions that result in tumor formation.

This integrated tumor- inducing DNA also directs the synthesis of an uncommon group of amino acid derivatives (opines) that only the bacterium can use as a source of carbon and nitrogen for further growth. The tumor- inducing DNA can be made nonpathogenic by removing the elements responsible for releasing the controls of cell division and for opine formation. The nonpathogenic DNA (T-DNA), which no longer induces tumor formation, can then be used to carry a different organism's gene into a host-cell chromosome.

As with the free DNA method, cells carrying T-DNA can be detected by incorporating selectable markers such as antibiotic or herbicide resistance. In this way, only cells carrying the resistance markers can grow on culture media in which the antibiotic or herbicide is incorporated; all untransformed cells are killed. The use of A. tumefaciens greatly increases the precision of DNA insertion.

Agrobacterium uses specific DNA-signaling sequences (T-DNA borders) to determine the start and stop points of DNA transfer to plant cells. Although there can still be substantial variation in the transferred DNA, the endpoints of DNA transfer are usually localized to a small region, within 10-100 bases. Moreover, the number of copies of inserted genes can usually be limited to one or a few. Recent improvements in transformation procedures have permitted researchers to largely switch from the free DNA techniques to Agrobacterium.

In any case, the precision of rDNA biotechnology permits accurate determination of the location and number of copies of the inserted DNA, even if the location of DNA insertion cannot be controlled. Scientific knowledge of the structure of the plant genome has grown as a result of research on the "laboratory plant" Arabidopsis thaliana, a small plant in the cabbage family that has only five chromosomes and grows from seed to seed in about seven weeks. Sequencing the entire genome of this plant is now almost complete.

Because of the great similarities among plants in general, Arabidopsis can be used as a crop plant analog, and DNA sequences from Arabidopsis of known function can be used to identify their homologs in economic crops. DNA markers can be used to identify chromosome regions that carry blocks of genes of individually small effect, quantitative trait loci or QTLs, which contribute to characteristics such as yield, maturity, baking quality, flavor, and aroma, making possible much more sophisticated selection procedures for plant breeding.

The opportunity to select and multiply a gene of interest and then introduce it into a crop plant was of great interest to most plant breeders because it heralded the era of directed genetic change. It was now possible to introduce a new gene into an accepted and adapted variety in a single step. This reduced the long and tedious process of winnowing out the many forms that are inferior to the adapted varieties that are characteristic products of most conventional breeding programs which introduce new characters from wild unadapted material.

In practice, rDNA biotechnology-derived forms can be better thought of as new forms of germplasm to be incorporated into breeding programs, thereby extending the range of characteristics available to a breeder. The breeder must still test the results to ensure that the step of introducing the non-host gene, or transgene, causes no other changes that would be detrimental to the farmer, the consumer, or the environment. These tests include detailed analyses of the composition of the product harvested from the rDNA biotechnology-derived form.

The first rDNA biotechnology-derived food plant marketed in the U.S. was the FlavrSavrTM tomato, introduced in 1994. Produced using T-DNA, this tomato carried an antisense gene for the enzyme polygalacturonase (PG), an enzyme formed as the fruit ripens and which is responsible, in large part, for fruit softening. The gene encoding PG was isolated, inverted in the cloning vector (producing an antisense form), and then introduced into cells that also carry the gene in the normal orientation. In the inverted DNA, the mRNA is transcribed from the wrong DNA strand to form an antisense message.

As a result, much less of the enzyme is produced. It was expected that the fruits of the tomato would have an extended shelf life, since they would not soften as rapidly as normal fruit. In fact, the FlavrSavr tomato was not a commercial success as a retail product because of uncompetitive agronomic characteristics; however, a processing variety engineered with a related construct proved to be useful to processors, since the ripe fruit has a higher solids content, resulting in economic and quality advantages.

Following the introduction of the rDNA biotechnology-derived tomato in 1994, other rDNA biotechnology-derived crops that contained modified agronomic traits soon followed. These plants included squash that are resistant to some strains of zucchini yellows and watermelon mosaic viruses in 1994, insect-resistant potato and cotton in 1995 and corn in 1996, and herbicide-tolerant soybean and canola in 1996. Although the consumer's awareness is largely limited to these products, there are many others under development that are expected to appeal more directly to consumers. These include fruits, root and leaf vegetables, and grains with enhanced nutritional and health-promoting properties.

RECOMBINANT DNA BIOTECHNOLOGY-DERIVED FOODS

Recombinant DNA biotechnology-derived foods are part of the continuing sequence of genetic improvement of the food supply.

Although it is sometimes portrayed as fundamentally new, the newness of rDNA biotechnology is best considered from a historical perspective. The plants and animals that modern agriculture produces today to feed the world's people are the result of more than 10,000 years of genetic modification and refinement. For example, there is the agricultural green revolution, which has contributed to increased human longevity and improved quality of life in developing countries.

The green revolution is viewed by many knowledgeable scientists as the latest major achievement in a long quest begun by ancient agriculturists who first cultivated and domesticated wild plants for food and fiber. Genetic modification of plants began approximately 10,000 years ago when man first used what is referred to as selective breeding. This technique simply involved saving seeds from the most vigorous plants in an environment for replanting at a later time. Over a period of many years, this selection resulted in higher-yielding varieties of the crop.

It is this type of selection that, for example, turned the wild precursor of modern maize, teosinte, into an important human food and animal feed crop in America. The same processes in the Near East-the Fertile Crescent-resulted in einkorn and emmer wheat, barley, lentil, pea, chickpea, and bitter vetch. Likewise, the progenitor of the modern tomato bears almost no resemblance to its modern relatives, which are the result of centuries of selection and DNA recombination at the organism level.

Selective breeding relies principally on sexually transmitted genetic diversity in a starting population. By picking the best or most vigorous plants, breeders over time enrich the genetic makeup of a plant for attributes such as higher yields, increased resistance to pests, and greater compatibility with production schemes. It should be noted that this process in itself runs counter to natural selection. Breeding involves selection for optimal growth for human purposes or other characteristics in an agricultural setting and in many cases is inconsistent with nature and the ability of the organism to survive under evolutionary pressure.

Therefore, human intervention has involved what can be called a primitive type of genetic engineering from the outset. An excellent example of breeding versus natural selection can be gleaned from the history of cultivated wheat. The seeds of wild wheat relatives are dispersed by the shattering of brittle seed heads. In the earliest stages of domestication, 10,000 years ago, forms that do not shatter were selected, which enabled gatherers to collect the ripe seeds rather than pick them up from the ground.

Such a mutation in nature would prevent seed dispersal and lead to rapid extinction of those plants in the wild. As the available unused genetic diversity of the species diminishes, the potential for improvement also decreases. Since crop improvement relies on genetic diversity, i.e., new sources of genes and expression of existing genes, continued improvement has required and will continue to require even greater diversity. This need for diversity led to the next developments in plant breeding when farmers discovered that crosses between certain closely related species would produce fertile offspring.

Cross-breeding (also known as interspecies or intergeneric breeding), either fortùitous or intentional, permitted recombination and selection among genes at a whole new level to provide new sources of genetic diversity and desirable traits. Interspecies or cross-breeding offers two possible outcomes. First, new species that contain all of the genes from multiple parents can be created. Thus, triticale, a fertile wheat-rye hybrid, became a reality.

The first wheat-rye hybrid plants, were completely sterile, but fifteen years later fertile sectors were reported on a spike that resulted from spontaneous chromosome doubling. Second, another alternative involves recombination, where a single genome is maintained in the offspring, but that genome now consists of randomly chosen copies of genes from either of the parent species. This latter type of breeding in a sense is the precursor to modern rDNA biotechnology; however, it is highly imprecise.

Large segments of chromosomes containing thousands of individual genes have been introduced from one species into another in this way. This type of technology is employed today by breeders of many crops, including tomato, soybean, canola, and cotton, which are all products of extensive genetic modification and selection. The products of naturally occurring interspecies crosses have been employed for thousands of years, and many of the foods eaten today are derived from such crosses. A good example is cultivated hexaploid wheat, which has three different genomes, each derived from a wild ancestral species.

For thousands of years, this technology has relied upon the ability of a genetic cross to produce fertile offspring. Thus, it is considered "natural." Many interspecific hybrids are infertile; for example, the original wheat-rye hybrids were sterile, and seeds could only be produced after spontaneous chromosome doubling had taken place. Thus, while interspecific crosses opened up a vast new genetic resource to plant breeders, the need for fertile progeny limited the usefulness of this diversity. Sometimes, a cross of two species can produce a viable embryo, which develops for a period of time, then degenerates and dies.

However, by using the technique known as embryo rescue, the embryo can be recovered shortly after fertilization and placed in an in-vitro tissue culture system. In this artificial setting, the embryo can develop into a mature, fertile plant. Tissue culture can thus expand access to genetic diversity by saving crosses that would not survive outside a laboratory. Some attention has been paid to the use of ionizing radiation and chemicals to induce mutations and expand the range of variation available to breeders, but very few successful new forms of crop plants have been obtained in this way.

The same is true of somaclonal variation arising in tissue culture. However, spontaneous mutations have been important in the development of some cultivated plants. All of these conventional techniques for crop improvement share the disadvantage that they are, by nature, imprecise and unpredictable and only occasionally useful. Spontaneous and induced mutation can lead to one desirable change and many undesirable collateral changes in an organism's DNA makeup, which must be selected out.

Breeders cannot and do not attempt to define in molecular terms the changes that they make within a genome. Rather, they employ standard selection procedures to screen for new plants with novel alterations and incorporate these plants into their breeding programs. In spite of the undefined nature of these changes, many years of experience have affirmed the safety and usefulness of genetically improved varieties. Plant breeders, farmers, food manufacturers, and consumers all have routine, frequent, and extensive exposure to these genetically improved varieties.

An excellent example of how breeders use all of the above techniques is the tomato. The tomato, Lycopersicon esculentum var. cerasiforme, originates from central Mexico. The original species bears little resemblance to current varieties, which are the result of much genetic manipulation. The growth habits of the plant, resistance to viruses, diseases, and nematodes, as well as fruit taste and appearance are a consequence of mutation, hybridization, and selection.

For example, resistances to several diseases, tobacco mosaic virus, and nematodes were introduced from the distantly related species, Lycopersicon peruvianum and Lycopersicon chilense. Crosses between these two species and L. esculentum required embryo rescue. Each new resistance represents the introduction of a large chromosome segment from the distant relative into L. esculentum. The typical introduced nonhost DNA segment contains between 100 and 1,000 genes. A specific example illustrates the imprecision of traditional breeding. Introduction of resistance to the fungal disease Fusarium crown rot involved a cross between an irradiated L. esculentum variety and L. peruvianum.

From this cross, a resistant plant was selected and used in subsequent breeding. This resistance gene, along with its complement of other genes, is present in many commercial varieties of tomato today. As the tomato is a member of the nightshade family and many of its wild relatives contain high levels of toxicants in the interspecific crosses with L. esculentum, breeders have selected for varieties with minimal toxicant content.

While there is no requirement for toxicant screening in traditional tomato breeding programs, it is widely practiced. Moreover, toxicant screening is an integral part of assessing the safety of new rDNA biotechnologyderived varieties. It is against this experience base that rDNA biotechnology must be examined and compared. Recombinant DNA techniques involve the introduction of one or a few defined genes into a plant. While these introduced genes are often from other, non-host sources, the introduction of non-host DNA is not novel.

In fact, remnants of an ancient Agrobacterium transformation have been identified in Nicotiana species. It is important to note that it is the very same Agrobacterium that is now used widely by researchers to introduce genes into plants. Similarly, microorganisms have been used in food technology for thousands of years. As early as 6000 B.C., Sumerians and Babylonians used yeast to brew beer. Although the ancients knew nothing about microorganisms and could not knowingly culture them, they nevertheless systematically selected those with desirable fermentation characteristics to improve their food.

In modern times, the increasingly powerful science of genetics has been systematically applied to produce many valuable variants of yeast and bacteria. Recombinant DNA techniques have provided both an important new set of tools and access to a broader range of markets. They enable researchers seeking specific plant characteristics to precisely identify, characterize, enhance, and transfer the appropriate individual genes rather than uncontrolled and randomly assorted groups of genes, hoping the desired ones were included.

Researchers can now readily move selected and well-characterized genetic material from virtually any source in nature, greatly increasing the diversity of useful genes available for crop and microbe improvement. The long, continuous search for improved plants and the benefits of useful microorganisms is now increasingly based on the use of rDNA biotechnology techniques. Microorganisms are used in the production of foods, beverages, industrial detergents, antibiotics, organic solvents, vitamins, amino acids, polysaccharides, steroids, and vaccines.

Practical applications of pre-rDNA biotechnology include a variety of organisms used in pest control. Biological agents are also used as growth promoters for plants. Preparations containing the bacterium Rhizobium, which fixes atmospheric nitrogen, converting it into nitrogen-containing ions that are essential plant nutrients, have been sold in the U.S. since the late 19th century. As early as the mid-1980s, these pre-rDNA biotechnology products, together, had a value in excess of $100 billion annually.

Since the introduction of rDNA biotechnology, many of these microorganisms have been improved, such as those used to produce the enzyme chymosin necessary for cheese production. Some critics of rDNA biotechnology have taken the view that it represents a fundamental change from traditional techniques for the genetic modification of plants and microorganisms. The same physical and biological laws govern the response of organisms modified by modern molecular and cellular methods and those produced by classical methods.

The NRC went on to characterize rDNA biotechnology as part of a sequence of scientific advances that has extended over a 10,000-year period. A 1991 joint Food and Agriculture

Organization/World Health Organization consultation, addressing the question of the safety of rDNA biotechnology-derived foods, came to similar conclusions: Biotechnology has a long history of use in food production and processing. It represents a continuum embracing both traditional breeding techniques and the latest techniques based on molecular biology.

The newer biotechnological techniques, in particular, open up very great possibilities of rapidly improving the quantity and quality of food available. The use of these techniques does not result in food which is inherently less safe than that produced by conventional ones. A timeline that shows the increasing power of genetic modification over the past 12,000 years appears. Even though food derived from biotechnology in the broad sense is hardly new, some critics nevertheless have been concerned that rDNA biotechnology may result in different and dangerous organisms.

Considering that there are tens of thousands of the host organism's own genes, the introduction by precise techniques of one or a few additional, well-characterized genes does not create an organism that is more likely to be changed in gross physical properties or wholesomeness than an organism derived through a traditional breeding program. Indeed, because of the greater precision in selecting the desired trait, an adverse result is unlikely.

A corn plant with a newly inserted bacterial gene that confers increased resistance to the European corn borer (a commercially important insect predator) is still a corn plant. Likewise, a microorganism long used for food production is not altered in any fundamental way by the insertion of additional copies of a geneencoded rate-limiting enzyme. Consider whether genetic recombination, itself, is of concern. It has already been established that people have long engaged in the systematic improvement of domesticated microbes, plants, and animals.

But the impact and importance of these changes are much smaller than what occurs continuously in nature. Innumerable recombinations between related and unrelated organisms have occurred by several mechanisms. Sexual reproduction randomly combines genes from two parents in the offspring, which then has a unique set of genes to pass along to the next generation. In the gut, decomposing tissue, and infected wounds, bacteria take up naked mammalian DNA, albeit inefficiently, when they encounter disintegrating cells, and some of this DNA may be in- corporated into the bacterial genome, but there is no established evidence that this happens.

Over the past million years and longer, mammalian-bacterial genetic hybrids have appeared, been tested by competition within bacterial populations and by environmental stresses, and conserved or discarded by natural selection. Similar genetic recombination and hybridization also has been widespread among fungi, viruses, and plants. Evolutionary biology provides data relevant to the issue of the uniqueness of chimeric genes created by rDNA biotechnology. The sequencing of various genomes during the past decade has revealed that nature has been remarkably conservative about maintaining and using effective molecules as they evolved.

Similar protein sequences and biochemical pathways are found in different species, across genera, and even across phylogenetic kingdoms. The Escherichia coli genome, for example, contains gene sequences that are closely related to those in a wide spectrum of organisms, ranging from other bacteria to plants, insects, amphibia, birds, and humans. Another issue, conversion of a nonpathogen into a pathogen through limited genetic recombination, is best considered within the context of the nature of pathogenicity. This process is both complex and multifactorial.

Pathogenicity usually is not a trait produced by a single gene; however, the transfer of a single gene to an organism that has all the other necessary genes can make it pathogenic. Pathogenicity requires the coordinated activity of a set of genes that affect essential properties. A pathogen must possess three general characteristics, each of which involves multiple genes. First, pathogens must survive and be able to multiply or produce toxin in or upon host tissues or food sources. This necessitates an appropriate oxygen tension, pH, temperature, water activity, and nutritional milieu.

Pathogens must be able to adhere to specific surfaces on or in the host. Second, the pathogen must be able to resist or avoid the host's defense mechanisms for the period of time necessary to multiply to sufficient levels to cause disease. Third, the pathogen must be able to survive outside of the host and must be disseminated to new host organisms. The organism must be meticulously adapted to this pathogenic lifestyle. On the other hand, a mutation that interferes with a gene essential to any one of the three characteristics of a pathogen can eliminate pathogenicity.

It is worth noting that severe pathogenicity is even more dependent upon favorable conditions and is, therefore, much rarer in nature than mild pathogenicity. The probability of creating and commercializing an organism inadvertently capable of producing a medical or agricultural problem is therefore quite small. The expert panels are of the view that this probability is lower with rDNA biotechnology than with the more random, less targeted, and less predictable traditional methods of genetic modification.

In rDNA biotechnology-derived organisms, typically one, two, or three genes are being inserted. The genes, gene products, and their functions are known. This information guides scientists in determining which possible risks are relevant and need to be explored. In comparison, with traditional breeding, a large number of genes with unknown functions are involved, making it much more difficult to sort through the progeny and focus on the relevant risks involved.

Adverse outcomes accompanying genetic change have always been possible but are routinely intercepted during the usual, extensive testing that takes place in growth chambers, greenhouses, and the field. Whatever the technique used to craft a variety, it goes through extensive testing before being used commercially, particularly if the developer chooses to enter it into formal seed registration programs. In practice, the testing is even more extensive in the case of an rDNA biotechnology-derived variety.

Therefore, the expert panels are of the view that rDNA biotechnology has the potential to reduce still further the chance that any such mishap will occur. The field and chemical testing that accompany it-even more thorough than in traditional genetic modification- make such an unfavorable outcome even more unlikely. Genetic changes that make a plant more useful to humans usually have made the plant less "fit" and less able to survive in the wild.

REGULATORY SIGHT OVER rDNA BIOTECHNOLOGY

Regulatory oversight over rDNA biotechnology spans three major federal agencies: the Food and Drug Administration (FDA), the Environmental Protection Agency (EPA), and the U.S. Department of Agriculture (USDA). Jurisdiction over the varied rDNA biotechnology products is determined by their use, as has been the case for products made by traditional means. More than one agency may be involved in regulating different aspects of an rDNA biotechnology- derived product.

As the regulatory mandate varies, so does the nature of the agencies' risk assessment and management protocols. Planned Introductions of Biotechnology Products into the Environment." This notice described a risk-based, scientific approach to the oversight of planned introductions of rDNA biotechnology derived products into the environment, focusing on the characteristics of the product and the environment into which it is being introduced, not the process by which the product is created.

The fundamental approach by the federal government to the review and regulation of rDNA biotechnology-derived products has largely been through a processbased trigger to oversight. Crops and microbes produced using rDNA biotechnology have been consistently subjected to higher requirements and standards than those applied to similar products produced using traditional techniques. At this time, there is less experience with rDNA biotechnology-derived products, but that experience base is increasing substantially.

Food and Drug Administration

FDA regulates different aspects of rDNA biotechnology under the authority of the Federal Food, Drug, and Cosmetic Act (FFDCA) and the Public Health Service Act (PHSA). FDA has a mandate to ensure the safety of all food sold in the U.S., as well as the safety and efficacy of pharmaceutical products. To date, FDA has conducted almost fifty reviews of rDNA biotechnology-derived plant products used for human food or animal feed.

Human Food and Animal Feed. Except for meat and poultry products regulated by USDA, FDA is responsible for ensuring the safety and proper labeling of food products for human consumption. FDA also regulates the safety and labeling of animal feed, taking into account both the safety to human consumers of animal-derived food products and the safety to the animal being fed. FDA's statutory authority is provided by the FFDCA. FDA has very broad authority to regulate the introduction of new food crops, whether conventionally grown, produced through hybridization or cross-breeding, or produced using rDNA biotechnology.

Every firm or individual that produces whole foods or food ingredients is legally required to ensure the safety of foods and food ingredients introduced into commerce. FDA has a number of enforcement tools that can be used to ensure the safety of food. Specifically, the FFDCA prohibits the adulteration of any food item that moves in interstate commerce. Of particular importance, foods are deemed adulterated if they contain certain poisonous and deleterious substances.

With certain exceptions that are not relevant to this discussion, the FFDCA defines a "food additive" as any substance, not "generally recognized as safe" (GRAS) by qualified experts for its intended use, that becomes a component or otherwise affects the characteristics of food. Food additives must be the subject of a petition to FDA, followed by FDA premarket approval; their manufacturers have the burden of establishing, through scientific testing, the safety of the substances.

In comparison, a food manufacturer that believes its food ingredient is GRAS may market the ingredient without seeking FDA's concurrence, subject to the risk that FDA will disagree and take legal action to remove the ingredient from the marketplace. In the U.S., whole foods such as fruits, vegetables, and grains are not regulated as "food additives" and are not required to undergo premarket approval; nor are they commonly subjected to extensive safety testing.

Thus, new varieties of crop plants produced by traditional breeding methods are not subject to FDA premarket review. Nevertheless, authority exists to ensure that such foods do not present a reasonable possibility that consumers might be injured by consuming them. With respect to all foods, FDA can initiate legal action to remove a food from the market if it is judged to present a health risk.

While there is no evidence that such authority has ever needed to be exercised with respect to traditional breeding practices, plant breeders and food processors have several times intercepted toxic food plants before they reached the market. FDA stated:

> This policy statement is a clarification of FDA's interpretation of the Federal Food, Drug, and Cosmetic Act (the act) with respect to technologies to produce foods, and

> reflects FDA's current judgement based on new plant varieties now under development in agricultural research. This action is being taken to ensure that relevant scientific, safety, and regulatory issues are resolved prior to the introduction of such products into the marketplace.

FDA set forth its authority to control food products derived by rDNA biotechnology techniques and listed the safety issues that need to be addressed in assessing the safety of whole foods that contain or use rDNA biotechnology-derived plants and microorganisms. One key point is that under certain conditions, foods and food ingredients derived from rDNA biotechnology-derived plants or microorganisms may be subject to the provisions of existing requirements governing food additives and GRAS substances.

FDA noted that in the case of foods derived from new plant varieties, it is the transferred genetic material and intended expression product(s) that could be subject to food additive requirements if these materials are not GRAS. FDA stated that if the intended expression product is a protein, carbohydrate, or other substance that differs substantially from substances currently present in food, then that substance might not be GRAS and may be a food additive requiring premarket approval.

Another important point is that if an rDNA biotechnology-derived plant or microorganism is used to produce a GRAS substance or an approved food additive, the resulting material would continue to be regulated in a similar fashion to the way in which it has historically been regulated. FDA's 1992 policy on new plant varieties applies irrespective of whether the plant arose from rDNA biotechnology or "conventional" genetic modification methods.

FDA does not routinely subject foods from new plant varieties to a premarket approval process or to extensive scientific safety tests. FDA's policy does, however, define certain safety-related characteristics of new foods-such as transfer of an allergen or increased levels of a natural toxicant-that trigger additional scrutiny. FDA's policy includes a flow chart for guidance that asks a series of questions directed to scientific issues of safety and nutrition of the foods derived from the new plant variety. The assessment focuses on the following risk-based considerations:

— Toxicants known to be characteristic of the host and donor species.

— The potential that food allergens will be transferred from one food source to another.

— The concentration and bioavailability of important nutrients for which a food crop is ordinarily consumed.

— The safety and nutritional value of newly introduced proteins.

— The identity, composition, and nutritional value of modified carbohydrates, fats, or oils.

Fundamentally, FDA's current policy is that existing requirements mandate the same safety standards for foods, food ingredients, and feeds, regardless of the techniques used in their production and manufacture. Nevertheless, FDA has maintained a "voluntary consultation procedure," in which producers of rDNA biotechnology- derived foods are asked to consult with the agency before marketing their products, and without exception they have done so.

Pharmaceuticals and Human Vaccines: FDA regulates rDNA biotechnology- derived pharmaceutical products for human and animal use under the FFDCA and the PHSA. FDA also regulates rDNA biotechnology-derived vaccines for human use under the PHSA, while USDA regulates vaccines for animal use. Under both the FFDCA and the PHSA, new products must be the subject of premarket approval, based on laboratory and clinical testing to show the safety and effectiveness of the products for their intended uses.

Two USDA agencies are relevant to the regulation of foods and other products derived using rDNA biotechnology.

Foods: The Food Safety and Inspection Service (FSIS) is responsible for regulating the safety and labeling of meat and poultry products for human consumption. FSIS consults with FDA regarding the safety of food ingredients. The Animal and Plant Health Inspection Service (APHIS) is the agency within the USDA charged with protecting American agriculture against pests and diseases. Under the Plant Quarantine Act and the Federal Plant Pest Act, APHIS can regulate the importation and interstate movement of plants and plant products that may result in the entry into the U.S. of injurious plant diseases or insect pests.

The field-testing and the commercial sale of agricultural rDNA biotechnology- derived crops are regulated by APHIS through a permit and notification system. USDA's regulations cover the introduction of organisms and products altered or produced through genetic engineering which are plant pests or for which there is reason to believe are plant pests. "Plant pests" include agents that can directly or indirectly injure or cause disease or damage in or to any plant.

A "regulated article" includes any organism or any product, which has been altered or produced through rDNA biotechnology, which is a plant pest, or for which there is reason to believe is a plant pest. The permit and notification system does not apply to plants that are modified through traditional breeding methods. Thus, USDA's regulatory protocol is process based.

The regulation is intended to prevent the introduction, dissemination and establishment of plant pests in the U.S. APHIS will grant a permit only if it determines that the plant poses no significant risk to other plants in the environment and is as safe to use as more traditional varieties. APHIS can authorize nonregulated status for an article through a petition for a "determination of nonregulated status." Nonregulated status allows a plant to be treated like any other plant, i.e., allows for the plant to be widely grown and commercialized.

Animal Vaccines: APHIS regulates animal vaccines under the Virus-Serum-Toxin Act. In general, animal vaccines are subject to premarket approval, based on testing to show their safety and effectiveness.

Environmental Protection Agency

EPA's stated mission is to protect human health and to safeguard the natural environment-air, water, and land- upon which life depends. EPA's responsibilities under the Federal Insecticide, Fungicide, and Rodenticide Act for registering pesticides, setting environmental tolerances for pesticides, and establishing exemptions for pesticide residues in and on crops are relevant to rDNA biotechnology-derived foods. A pesticide is any substance or mixture of substances intended for preventing, destroying, repelling, or mitigating any pest.

The standard measures the aggregate risk from dietary exposure and other non-occupational sources of exposure. EPA must now focus explicitly on exposures and risks to infants and children, assuming when appropriate, an additional safety factor to account for uncertainty in data. If EPA determines that there is a "reasonable certainty that no harm" to the public will result from aggregate exposure to a particular pesticide residue, then that residue level will be deemed "safe."

In the case of pesticides produced by plants developed using rDNA biotechnology, EPA's November 23, 1994, proposed rule takes the view that its regulatory process is focused on the pesticide and not on the plant; plants are subject to regulation only if they produce plant pesticidal proteins as a result of modification with rDNA techniques. Although EPA has not finalized that proposed rule, EPA has been implementing its essential elements since 1995. EPA's evaluation of products of rDNA biotechnology is distinct from the procedures used to assess the safety of the products of more conventional technology.

CHAPTER 4

BENEFITS OF GENETICALLY MODIFIED FOODS

In the harsh world of nature, surviving plants (and animals) have evolved to resist environmental stresses and both pest and pathogen attack. The survival of crop plants (and livestock) has been enhanced by human selection and intervention to produce food. In little more than a century, starting with hybridization, which was commercialized in the first decade of the 20th century, scientific breakthroughs enabled new types of plants, such as triticale and seedless watermelons and grapes, to be produced.

Starting in the mid-1970s, genetic modification or "engineering" by molecular means became feasible. A healthy plant produces a variety of compounds to protect itself from being eaten or destroyed. At the levels commonly consumed in food, few of these naturally occurring compounds are deleterious to human health. However, many of these substances are hazardous if sufficient quantities of the plant are consumed under certain circumstances.

Examples include glycoalkaloids in potatoes, cyanogenic glycosides in cassava, trypsin inhibitors in lima beans, and allergenic proteins in a variety of foods. Plant breeding often has been successful in producing plants with increased pest and disease resistance, while retaining high yields and both taste and processing attributes. Synthetic pesticides are frequently used to produce high-quality and economically viable crops, such as apples and squash.

Food crops can be devastated by both above- and below ground microorganisms. While some compounds are available for combating fungi, such as the copper sprays and sulfur used by organic farmers and fungicides used by farmers and home gardeners, these are high-cost and broad-spectrum, often killing beneficial organisms as well. Except for plant breeding and use of insecticides for killing insects that transmit disease, no remedies for combating plant viruses are known. Unlike combating human bacterial infections, very limited remedies are available for preventing bacterial diseases of plants.

Antibiotics are available, but their use is not economical. The susceptibility of a plant to biotic and environmental stresses-such as temperature extremes, chemical challenge and heavy metal exposure (e.g., selenium), and salinity and drought-are affected by the plant's genetic composition and structure. For example, some leaves have evolved to conserve moisture or resist heat or freezing. Breeders have changed leaf and stem architecture to capture more sunlight and allow for greater air flow through the leaf canopy.

All plants with genes for resistance to pests and disease are evaluated by breeders and processors during initial stages, and are either subsequently commercialized as generally recognized as safe (GRAS), or, if changed by molecular means, reviewed by federal agencies before commercialization. Numerous scientific associations have become valued participants in the development and regulatory processes for these crops and food products.

Further, AACC has led the advancement in functional criteria for cereal grains such as the baking quality of wheat, relationships of the protein content of corn to product attributes, and other properties important or essential to various food products. This process may take up to 15 years from initial seed selection. Yield is very critical, as are processing properties, quality, composition, and organoleptic properties. rDNA biotechnology offers the potential for enhancing plant availability and survival, as well as growth.

For example, a severe strain of papaya ringspot virus in Hawaii threatened to kill the trees and decimate the livelihood of growers. Little resistance was available for breeding potential. Hence, the viral coat protein gene was transformed into stock, allowing the trees to grow. In 2005, methyl bromide, a soil fumigant widely used in certain areas, must be taken off the market, as a result of international treaty. At present, there is no alternative for controlling soilborne fungal pathogens in strawberry varieties grown in those regions; rDNA biotechnology offers the potential to retain the availability, at reasonable cost, of strawberries.

Apple and pear production is constrained by the bacterial disease called fireblight, first described in the 1870s. No satisfactory antibacterial compounds or adequate resistance to the disease is available in apples desired by consumers. rDNA biotechnology research has produced the first trees to resist this devastating disease. Grape vines, which require multiple years to grow and mature before production of both wine and table grapes, are subject to fungi, insects, root disease, and pest problems, which are increasingly difficult to control. Both the quality and availability of wine and whole grapes can be affected. rDNA biotechnology again offers the potential for minimizing damage caused by these agents.

The advantages of rDNA biotechnology-derived food crops-increased yield and better resistance to pests, disease, and environmental stress-are clearly apparent to growers but not to most consumers. Widespread acceptance of rDNA biotechnology-derived crops will not occur until consumers become convinced of their advantages. Advantages are consumers likely to derive from such crops are:

— Food-deficient regions of the world may becoine less common. rDNA biotechnology- derived crops can be developed to prosper under conditions that previously limited or prevented plant growth. This approach will increase world crop production and also increase the variety of crops suitable for growth in any given area. It is unlikely that an abundance of crops of this type will ever become available in a reasonable time frame through conventional breeding practices. Increases in the food supply that are potentially achievable by rDNA biotechnology are likely to greatly exceed those accomplished during the Green Revolution, which relied on conventional breeding practices.

— Improvements in the organoleptic and nutritional quality of foods derived from plants will occur more rapidly and be more pronounced through rDNA biotechnology modification than by conventional breeding.

— Improvements in the shelf life of fresh fruits and vegetables that either cannot be obtained through conventional breeding or are obtained only at a much slower rate will be attainable through rDNA biotechnology modification.

— Reduction in crops of the types and concentrations of allergens, naturally occurring toxicants, and other undesirable constituents will be more easily achieved by rDNA biotechnology modification than by conventional breeding.

— Introduction into crops of diseaseresisting and health-promoting constituents which would be exceedingly difficult, slow, or more likely impossible by conventional breeding will be possible by rDNA biotechnology modification.

Better Food

Recently, the human population of the globe passed 6 billion, and forecasts predict that this number will grow to 9 billion by

2050. While these numbers are more modest than the prediction only five years ago that the population would double by 2030, demographers predict that the vast majority of the growth will occur in Asia, Southeast Asia, and Africa, areas already under significant strain for food production. Even though improved agricultural practices and higher-yielding crops will likely meet the minimum number of calories to sustain human life globally, there is real and significant concern that the needs for adequate nutrition will not be met.

For example, although India produces sufficient food to prevent starvation, more than 30% of its population is malnourished. The situation is even more pressing in Africa, where diseases such as AIDS have reduced the numbers of farming women and children. Furthermore, periodic famines in arid regions in Africa continue to drive increasing numbers of people to malnutrition and starvation.

An infant born of a malnourished mother carries the effects of malnutrition into the fourth generation beyond the mother. The challenge is not simply to provide a steady supply of food, but a nutritious and safe food supply that improves the health and productivity of the global population. The past 10 years have seen the development of rDNA biotechnology that can play an important, but certainly not the sole, role in increasing the supply and quality of foods for peoples in the developing economies.

Despite the modest "farm surpluses" currently being produced in some areas (e.g., North America, Australia), the world does not yet grow nearly enough food to meet the demands of the 21st century. Even with human population stability expected by 2050, the world will need farm outputs that are 2.5-3 times greater than current harvests to provide high-quality diets to the world population just five decades from now. Technologies such as hybrid seeds, irrigation, nitrogen fertilizer, and integrated pest management are already in broad use on the world's best farmlands.

Extending the known non-rDNA biotechnology farming systems to the world's low-yield farming sectors might not even double current world farm output. rDNA biotechnology is the most important unused technology available to meet this last, huge surge in global farm demand. It is essential that we improve food production and distribution to feed and free from hunger a growing world population, while reducing environmental impacts and providing productive employment in low-income areas.

This will require a proper and responsible utilization of scientific discoveries and new technologies. The developers and overseers of rDNA biotechnology applied to plants and microorganisms should make sure that their efforts address such needs. The academies stated that foods can be produced through the use of rDNA biotechnology that are more nutritious, stable in storage, and in principle health promoting-bringing benefits to consumers in both industrialized and developing nations.

Any shortfall in the effort to triple current yields on the world's existing farmlands over the next 50 years is likely to mean massive malnutrition for the world's poorest peoples. It is also likely to mean the plowdown of millions of square miles of forests and wildlands and thus the probable sacrifice of millions of irreplaceable wild species. Organic farming is often held up as the agricultural ideal for the 21st century. Unfortunately, the U.S. has only about one-third of the organic nitrogen needed to support current U.S. farm output.

Countries such as India and China have even less of the organic nitrogen that would be needed; they already feed much of their biomass to livestock and burn animal feces for cooking. The world as a whole may have only one-fourth of the organic nitrogen necessary to support its current food production, let alone triple for 2050. In addition to sheer caloric needs, hundreds of millions of the world's poor people are still short of the protein and micronutrients needed to ensure long and healthy lives. rDNA biotechnology has already demonstrated, through such successes as chymosin, "golden rice," and acid-tolerant crops, that it is one of the most promising ways to meet these urgent needs.

"Golden rice" has been genetically modified through rDNA biotechnology to have increased beta-carotene content, which may help to overcome the severe vitamin A deficiencies which cause millions of poor children to go blind or die every year in low-income, rice-consuming cultures. A related product of rDNA biotechnology may also help eliminate the iron deficiency which threatens hundreds of millions of rice-culture women and their babies with birth complications each year.

Toxic metals, such as aluminum and manganese, are widely present in "acidic" tropical soils, which account for nearly half the arable land in the tropics. These metals reduce root growth, cutting yields by up to 80%. To produce acid-tolerant crops, two researchers in Mexico inserted a gene from a bacterium into tobacco and papaya. The plants thus secrete citric acid from their roots, chelating these toxic metals. The yield gains now anticipated from making such soils accessible will be critical to protecting the tropical forests, which contain most of the world's species of plants and animals.

rDNA biotechnology should also be able to play key roles in protecting, preserving, and processing foods, to minimize food losses, maintain or improve quality, and increase processing efficiency. The result should be better health, greater food enjoyment, and still less competition between people and wildlife for scarce land. If the need for improved farming and food technologies is viewed in purely economic terms, the 21st century will see a huge increase in the demand for farm exports.

The rapidly rising affluence and demonstrated protein hunger in such densely populated countries as China, India, and Indonesia virtually guarantee that the temperate-zone countries well endowed with farmland will have the opportunity to help meet the soaring food demand in the emerging economies. Agriculture in the developed countries thus could make a major environmental contribution while increasing production each year for additional farm exports. Hundreds of thousands of additional urban jobs would also be created by such an expansion of farm exports, in food processing, transportation, and many associated fields.

FOOD TECHNOLOGY AND BIOPROCESSING BENEFITS

Production of various foods and food ingredients through fermentation, also called bioprocessing, has occurred since the earliest records of man's preservation of foods. Microorganisms and enzymes are used widely for the conversion of raw food substrates (e.g., milk, cereals, vegetables, and meats) into a plethora of fermented products (e.g., cheeses, cultured milks, sourdough bread, pickles, wine, beer, and sausages).

Bioprocessing technology has been further developed for specialized production of food ingredients (e.g., organic acids, amino acids, vitamins, and gums), or processing aids (enzymes). rDNA biotechnology is expected to revolutionize food bioprocessing across all of these arenas through improvements in the responsible microorganisms (e.g., bacteria, yeasts, and molds) and efficient production of specialized enzymes and ingredients via fermentation technology.

Ingredients made by bioprocessing are among some of the most attractive products of rDNA biotechnology. An understanding of metabolism and the ability to redirect metabolic pathways provide opportunities to produce food ingredients of higher quality and purity, as well as new ingredients for purification or synthesis that are not available through conventional methods. As discussed below, key ingredients targeted by rDNA biotechnology include organic acids, bacteriocin preservatives, enzymes, and microorganisms used for processing aids. rDNA biotechnology is also important in the production of vitamins and amino acids.

Organic Acids: Organic acids are commonly used as food acidulants and are among the most versatile food and beverage ingredients because of their solubility, hygroscopic, buffering, and chelation properties. Lactic, citric, gluconic, and propionic acids are all naturally occurring and produced by fermentation. Citric acid has the widest range of applications and with acetic acid accounts for 75% of food acidulant usage. Citric acid is produced by several molds, yeasts, and bacteria by fermentation of glucose via glycolysis.

Mutants of fermentation strains have been selected with steps in the Krebs cycle blocked to maximize accumulation of citric acid. Improvements via rDNA biotechnology have increased the rate of glucose fermentation and eliminated enzymes that degrade citrate in the production organisms. Lactic acid is used as an acidulant for cheeses, meats, jellies, and beer. Derivatives of ammonium lactate are used as sources of non-protein nitrogen in animal feeds, and sodium or calcium stearoyl lactylates are used as emulsifiers and dough conditioners. Fermentation processes produce both the D and L isomers of lactate via the two stereospecific lactic dehydrogenases-L-LDH and DLDH.

L-lactate is the natural and preferred form for food use because it is the form used by humans and the D-form is considered slightly toxic. Two key improvements have occurred via rDNA biotechnology. First, elimination of D-LDH by gene replacement leads to pure L-lactate production in Lactobacillus species. Second, the bovine L-LDH gene was introduced into Kluyveromyces lactis, leading to significant yield increases in L-lactate. Lactic acid bacteria show considerable promise for metabolic engineering because their biosynthetic pathways are completely separate from their energygenerating pathways.

Bacteriocin Preservatives: Bacteriocins are peptide antimicrobials that kill bacteria. Nisin is notable among these because of its broad killing range against Gram-positive pathogens and its GRAS status based on its safe consumption in dairy products for centuries. Genetic approaches to understanding the regulation of nisin biosynthesis in Lactococcus have identified the fermentation conditions where nisin, or other enzymes/proteins, can be overproduced, approaching 50% of the cells' protein. The increased availability of nisin has led to expanded applications for this preservative in foods. Moreover, expression systems using the nisin-inducible promoter are already providing powerful tools for production of food-grade enzymes and protein ingredients by L. lactis.

Enzymes: Enzymes were important agents in food production (e.g., milk clotting, bread production, juice clarification, alcoholic beverage production) long before modern rDNA biotechnology was developed. Today, enzymes are indispensable to modern food processing technology. The U.S. market for enzymes used in food manufacture is expected to grow to $214 million by 2006. An increasing variety of food enzymes has been produced using rDNA biotechnology.

Their accepted use in foods is based on the following facts: enzymes produced by rDNA biotechnology are identical to their natural counterparts (e.g., chymosin); enzyme preparations are free of any deleterious substances that could be introduced during the bioprocessing and purification steps; and viable rDNA biotechnology-derived microorganisms are not present in the final preparation.

Chymosin is the most important enzyme used in the dairy industry to clot milk. Its specific hydrolysis of kappa-casein destabilizes milk micelles and leads to rapid coagulation, clean flavor, and maximum protein yields from cheese curds. Traditionally, chymosin was obtained from rennet extracted from the stomachs of young calves. Rennet supplies faced major declines as calf slaughter decreased during a period of increasing worldwide cheese production. Several commercial entities undertook efforts to clone and express chymosin, in its exact natural form, from bacteria (E. coli), yeast, and molds.

Chymosin that was produced in bioreactors was identical to the animal-derived enzyme and was substantially purer (>95% chymosin) than traditional rennet (containing only 2% chymosin). Since these were the first products of rDNA biotechnology targeted for use in human foods, extraordinary precautions were taken to assure that the enzyme preparations were free of toxins, had no live recombinant organisms, and exhibited no ill effects in animal studies. Indeed, although E. coli is not a foodgrade organism, the Food and Drug Administration concluded that chymosin produced from recombinant E. coli was identical to its conventional counterpart and, therefore, could be considered to be a GRAS substance acceptable for use in foods.

Estimates of the use of rDNA biotechnology-derived chymosin now exceed 80% of the market in the U.S. and Canada, where cheese produced using rDNA biotechnology-derived chymosin is regarded as vegetarian, kosher, and halal. The chymosin example established the basis for production of a variety of safe and functional rDNA biotechnology- derived food-grade enzymes. Improvements are readily apparent in enzyme availability, purity, and cost, which benefit and improve the quality of foods available to consumers.

Because of the considerable benefits to be realized, it is probable that most food processing enzymes eventually will be rDNA biotechnology-derived. Enzymes of higher purity and specificity can be obtained, which will improve processing efficiencies and quality, while reducing energy costs, waste, and environmental impacts.

Processing Aids-Microorganisms: Many beneficial microorganisms are directly responsible for the preservation and processing of food, including primarily yeasts (Saccharomyces cerevisiae), beneficial molds, and lactic acid bacteria. rDNA biotechnology offers considerable promise for the beneficial modification of microorganisms that drive food fermentations.

In those examples where rDNA biotechnology modifications were made in yeast, self-cloning was used. This concept is based on the fact that DNA rearrangements occur naturally and often within the genome of any given organism. Self-cloning protocols require that only DNA originating in the host organism can be manipulated and reintroduced to create an improved microorganism. For bioprocessing microorganisms, there remain many attractive targets for improvement through rDNA biotechnology.

Most of these offer improved product quality, better control of fermentations, and thus enhanced food safety. It is pertinent to note that progress in molecular biology and genomics has provided both the tools and targets for precise genetic modification of microorganisms. Specific genes and DNA sequences can be introduced, eliminated, or altered in a precise manner in microorganisms in many cases. If desirable, marker genes and extraneous DNA can be removed.

These specific changes are superior to nondiscriminating strategies that have been used historically to mutagenize DNA, in chance efforts to select more-efficient organisms. History claims many successes in the selection of improved organisms (e.g., 10,000-fold increase in penicillin production from Alexander Fleming's original Penicillium strain). However, the mutations are random and nondiscriminating, and any effects of secondary mutations are not known. In contrast, rDNA biotechnology has provided the tools and information to make more-specific genetic changes, ensuring both performance and safety.

ANIMAL BENEFITS

Animal Feed: Over the past century, a remarkable revolution in both animal and plant agriculture has led to a relatively stable and cheap supply of food for the developed world. In parallel with the Green Revolution, there has emerged an equally important revolution in animal feed efficiency. This was essential to provide for the developed world's demands for high protein content of their diets.

This divergence between plant supply and animal demand has occurred because selection for improved plant productivity has focused primarily on plant yield per acre of land and has largely ignored the nutritional needs of the animal. In parallel with this, improvements in animal genetics have focused primarily on increased efficiency of feed conversion, assuming a near optimal feed composition. Thus, plant and animal breeders have been inadvertently widening the gap between the increasing nutritional needs of the animal and the ability of the crops used for feed to meet those nutritional needs.

To optimize animal nutrition, the animal feed industry has flourished during the last century. The growth in the feed supplements business was critical to meet the demand for macro- and micronutrients as feed additives. By monitoring essential feed components, leastcost feed formulations have evolved. Coupled with a vertically integrated system from farm to consumer, this has resulted in a supply of high-quality meat at affordable prices in the developed world. The basic commodity-feed components of calories and protein supplied by the crop plants are the base of the supply chain.

Today, a significant proportion of world crop production is directed into animal feed in the form of forage, silage, or grain. For example, 80% of U.S. maize production is used for feed for poultry, hogs, and cattle. Although the developed world's meat consumption is projected to be stable, world meat consumption is expected to rise very significantly during this century because of increased demand for animal products in the developing world.

By examining today's animal feed conversion ratios, it becomes clear that the greatest efficiencies in animal productivity are primarily based on chickens and hogs. For every pound of weight gained by a chicken, approximately 1.5 pounds of feed is required. For hogs, the feed requirement is approximately 4 pounds for every pound gained, whereas for beef cattle the feed requirement is more than 10 pounds. These conversion ratios demonstrate the importance of a high density of essential nutrients as well as calories.

The shortfall in calories and essential amino acids available from cereal feed is partially offset by mixing corn with soybean to make a mixed soybean/ corn feed. To improve the caloric value further, feed is supplemented with fats such as animal offal and feed-grade animal and vegetable fats which include by-products of the restaurant, soap, and refinery industries. Other nutritional needs are met by adding various feed additives to the mix. Productivity is enhanced by managing an optimal environment for the animals to grow. It is here where significant emphasis is placed on animal health as it relates to carcass quality.

The first major improvement in animal nutrition was the addition of vitamin D, which allowed chickens to be raised in controlled environments. This in turn minimized losses due to environmental changes, predators, and disease. Further improvements came with diets supplemented with vitamin E as an antioxidant, methionine to improve immune function, conjugated linoleic acids to improve feed efficiency and carcass quality, enzymes to improve digestion and remove antinutritional factors and toxins, antibiotics to optimize animal health and stabilize weight gain, prebiotics and probiotics to improve gut microflora, and growth hormones to improve feed efficiency.

Today, there are many feed supplements and additives with varying efficacies. Some are macronutrients and others micronutrients, yet others are more veterinary pharmaceutical in nature. Most are aimed at improving carcass quality while maintaining or improving animal feed efficiency. A wide array of methods for producing these feed additives are used across the world, from fermentation to synthetic chemistry, and some already rely on the application of rDNA biotechnology. The key emphasis is placed on quality linked to the cost of production. As we move forward in the new century, a new revolution is occurring.

Through dramatic advances in genetics and rDNA biotechnology, it is now possible to envision ways of enhancing animal feed by directing the plant itself to produce a more nutritious product. With this advance comes the opportunity to redesign and rethink the basic composition of feeds derived from silage, forage, and grain. This change will go beyond plant yield as a commodity product and enter the realm of value-added crops. In the developing countries, where meat consumption is very much lower than in the developed world, these rDNA biotechnology-driven advances in plant composition for animal feed present another opportunity to improve human nutrition.

Collectively, these genetic improvements in crop composition have been termed the "output traits" to distinguish them from the input traits that were the hallmark of the first wave of rDNA biotechnology- derived products, which included resistance to herbicides, insects, and viral diseases. Thus, a new industry to improve feed crops is emerging as an adjunct to the existing feed supplement industry. The seed genetics industry, linked to rDNA biotechnology, is altering seed and plant traits to improve basic plant components.

Some output traits already on the market include silage corn improved by the mutant brown-midrib trait, producing so-called BMR-corn. This corn has a mutation in the pathway leading to lignin deposition that significantly increases its digestibility for ruminant animals. A more recent improvement in maize was the development of corn with a higher-oil content.

In this case, the maize seed was selected to have a significant increase in oil content, from 3.5% in conventional maize to about 6% oil in higher-oil corn. Varieties having the higher oil content have grown significantly in market share over the last few years. A further improvement came with the development of corn with increased protein and essential amino acids necessary for optimal animal growth.

In soybeans, improvements such as altered oligosaccharides to reduce nondigestible components and increased oleic acid composition have emerged. Increased plant resistance to fungal infections will reduce the risk of serious contamination with mycotoxins. In all crops, it seems reasonable to expect additional improvements through further enhancements in oil, fatty acid, protein, starch, carbohydrate, vitamin, antioxidant, and mineral composition. A new area of animal feed improvement aids control of environmental pollution.

In the intensive livestock industry, there are significant problems with odors and controlling the release of unused nutrients, such as nitrogen and phosphorus from animal waste. A significant step forward in resolving the major political and economic issue of phosphorus pollution was the identification of approaches to control phytic acid content of feed by reducing the phytate content of seed via rDNA biotechnology or adding phytase to feed via supplementation. A new development involves reducing the phytate content of the seed significantly by introducing the lpa1-mutant of corn.

Low-phytate corn is new on the market. Recent studies have revealed that it has an unexpected nutritional enhancement, namely, an increased bioavailability of amino acids. Phytate also strongly chelates iron, calcium, zinc, and other divalent mineral ions, making them no longer bioavailable. This means less phosphorus waste as well as reductions in nitrogen waste. Collectively, these genetic enhancements in feed composition have been achieved by introducing valuable traits directly into the commodity component, the plant itself. This is a significant technical challenge for both conventional plant breeding and rDNA biotechnology.

Nevertheless, significant progress has already been made through a combination of conventional breeding, germplasm screening linked with high throughput tests of specific traits, rDNA biotechnology-aided breeding using DNA markers, selection for mutants carrying specific traits, and rDNA biotechnology (transgenic) approaches, which involve isolating, characterizing, and modifying individual genes followed by plant transformation and trait analysis. In all of these approaches there is an overriding imperative to maximize plant yield, because the least-costs feed analysis will continue to rule this market sector.

Thus, it will be necessary to provide a platform which will continue to be based on delivering products using the cheapest methods available. This is perhaps the most challenging step of all, since yield requires many genes all functioning optimally in the plant. These enhancements are a key next step in securing and improving the world food supply in the new century for both the developed and the developing nations. Furthermore, these enhancements are expected to also alleviate some environmental problems associated with intensive livestock production.

Plant-Based Animal Vaccines: Animal health benefits appear to be possible, as certain vaccines and growth hormones are amenable to an rDNA biotechnology approach. At present, animals receive multiple injections to maintain their optimal health and high feed efficiency. This is inconvenient, causes some distress to the animal, and can cause some meat spoilage. By engineering a plant to express some of these products it appears to be possible to circumvent these concerns. Further advances seem highly likely in this type of technology.

Howard demonstrated production in plants of a vaccine against transmissible gastroenteritis virus which protected swine in clinical trails against the virulent pathogen. Dalsgaard et al. demonstrated that a cowpea plant-based vaccine protected mink against the diarrhea and anorexia caused by mink enteritis virus (MEV), a member of a group of viruses that is also responsible for disease in cats and dogs. To develop the plant-based vaccine, a segment coding for the epitope of MEV was fused into the coat protein gene of a cowpea mosaic virus.

ENVIRONMENTAL BENEFITS

Farmers and producers have enthusiastically embraced the new varieties of rDNA biotechnology-derived crops that exhibit increased resistance to pests; tolerance to more environmentally benign herbicides (e.g., corn, cotton, soybeans); and virus resistance (e.g., squash, cucumbers, papaya). Although a major driving force for adoption of these crops is economic, farmers also welcome the environmental benefits of fewer pesticide residues and simplification of farming practices. In general, farmers who use the new varieties have realized significant savings in production costs, as well as increased yields.

These savings occurred despite the increased costs of seeds and "technology fees" that were added by seed producers to recover expenses for research and development. A recently released summary of the proceedings of the Ceres Conference on Agricultural Biotechnology describes the results of independent third-party studies to document farmer acceptance and profitability of rDNA biotechnology-derived crops.

In addition to economic advantages, the number of pesticide applications is typically reduced. Canola fields require only one herbicide application (instead of two), and glyphosate provides broader-spectrum weed control. Certainly, rDNA biotechnology can play an important role in development of agriculture that uses fewer and more-benign agrochemicals than needed with traditional crop varieties. Glyphosatetolerant soybeans offered easier weed management, less injury to crops, no restrictions on crop rotations, increase in no-till agriculture, and reduced costs.

U.S. farmers using glyphosate-tolerant soybeans saved an estimated $220 million in 1998 due to lower herbicide costs. The broad spectrum of weeds controlled by glyphosate means that soybean growers no longer need to make multiple applications of complex combinations of herbicides. Before the introduction of tolerance genes from other organisms, herbicides were selected by screening for chemicals that caused minimal crop damage while killing as many common target weeds as possible.

Broad-spectrum herbicides like glyphosate were most often used to kill vegetation in places like railroad tracks, paths, and parking lots. Glyphosate inhibits an enzyme essential for the synthesis of aromatic amino acids in plants. A form of the enzyme that carries out the same step in a bacterium and which is not inhibited by this herbicide. Glyphosate-tolerant soybeans carry the bacterial gene and are relatively insensitive to the herbicide. Extensive tests by the manufacturer of glyphosate have shown that it has a very low mammalian toxicity and is rapidly degraded in the soil after application.

Its microbial degradation ultimately produces carbon dioxide and water. No toxic intermediates or derivatives have been identified among its breakdown products. Unlike earlier herbicides that persisted in the environment and contaminated ground water, glyphosate appears to be safe and to disappear rapidly. Similarly, rDNA biotechnology is now being used to develop varieties of soybeans and other crops that are tolerant to other herbicides, which would otherwise kill them.

Two other herbicides with different modes of action, glufosinate and imidazolinone, are also being used on crops protected with transgenes. Corn resistant to sulfanyl urea has also been produced by the selection of mutants in tissue culture. The introduction of herbicide tolerance has contributed to the development of no-till agriculture and crop rotation with benefits that include savings on fossil fuel in preparing seedbeds, and reductions in soil erosion and in air pollution from burning crop residues.

Economic Benefits

The most widespread rDNA biotechnology- derived crops in the U.S. at the present time are cultivars of soybean, cotton, and corn. In the U.S. in 1999, 35% of the corn acreage was made up of either insecttolerant (23%) or herbicide-tolerant cultivars; 45% of the cotton acreage was insect tolerant; and 54% of the soybean acreage was herbicide tolerant. A USDA/ERS study found that herbicide-tolerant soybeans reduced farm input costs by 3-6% and increased average yields by more than 13-18% in most regions of the U.S. Estimated benefits of herbicide-tolerant corn and canola range from $15 to $24 per acre.

Considering that the planted area of rDNA biotechnology-derived crops more than doubled in 1998 to nearly 69 million acres, many farmers have obviously become convinced that rDNA biotechnology-derived varieties have superior characteristics. Certain segments of the commercial seed markets have already become highly concentrated. Two companies together account for more than 50% of North American sales of corn seed and nearly 40% of North American soybean seed sales.

Even in the relatively concentrated U.S. hybrid corn seed market, increases in seed costs have been less than half the value of yield increases attributable to new varieties for the period 1975 through 1998. The distribution of benefits resulting from introduction of Bt cotton estimated that the biotechnology firms involved captured 44% of the value of the innovation, with farmers receiving 48% and consumers 8%.

As competitors develop their own varieties of Bt cotton, the premium that can be charged for this variety can be expected to decline. The effect of competition can already be seen in the market for herbicide-tolerant soybeans. In this more-competitive market, the share of the economic benefits captured by the companies that introduced the first of these varieties is estimated at less than 25%.

HEALTH BENEFITS

rDNA biotechnology has the potential to improve the nutritional status of populations throughout the world. Both developed and developing societies can benefit from rDNA biotechnology-derived plants that will provide increased quantities of foods, as well as foods with unique and more-effective nutritional composition and qualities that will satisfy the individual needs of different populations.

There are many types of malnutrition, but all can be traced to two major sources, the lack of proper quantity and quality of foods. rDNA biotechnology offers unique opportunities to increase the quantity of food that is available in developing countries. In both developing and developed countries, rDNA biotechnology can also improve the nutritional quality of foods. Specific foods can be developed to correct malnutrition problems that are unique to different regions of the world. Plants can be modified to grow well in areas of low production potential.

They also can be modified to provide increased and more-stable quantities of essential amino acids, vitamins, or desirable fatty acids. For example, deficiencies of vitamin A and iron are serious, life-threatening health problems in many regions of the developing world. Vitamin A deficiency can increase susceptibility to infections and cause blindness.

An inadequate consumption of iron results in anemia. According to the World Health Organization, vitamin A deficiency affects approximately one quarter of a billion children, with child death rates as high as one out of four in some regions of the world. Iron deficiency affects 3.7 billion people. rDNA biotechnology-derived golden rice with increased content of beta-carotene, the precursor to vitamin A, is under development, and foods with enhanced iron content are also in the research pipeline.

Other regions of the world suffer malnutrition because their dietary sources of protein are inadequate. Because diets provide an inadequate source of protein, children suffer from stunted growth, increased susceptibility to infections, and impaired intellectual development. Through the use of rDNA biotechnology, the essential amino acids content of cereal grains such as corn and rice can potentially be increased to improve both the protein quality and quantity, thereby eliminating this form of malnutrition. Similar efforts to improve protein content and quality through conventional methods have met with only limited success.

As the science of nutrition improves and develops our understanding of the relationships between genetics, diet, and degenerative diseases, recommendations for dietary consumption practices will also change. In addition to classical nutrients, other plant components (i.e., phytochemicals), are now recognized for their contributions to improved health and the prevention of some degenerative diseases.

It will also be important to provide foods of appropriate composition to achieve maximum benefits. Scientists predict that in the near future rDNA biotechnology-derived foods with improved levels of phytochemicals and micronutrients will be developed. Some have predicted that these and other products will be well received by health-conscious consumers, who spend more than $6 billion annually on overthe-counter food supplements.

Probiotics are living microorganisms, typically delivered through foods, that offer benefits to health and wellbeing that are beyond basic nutrition, such as increased resistance to foodborne illness, decreased risk of some cancers, and potential lowering of blood cholesterol. Selected members within the Lactobacillus and Bifidobacterium genera are considered key probiotic species as they are able to survive stomach and intestinal transit, exert health benefits, and favorably affect the microbial ecosystem.

rDNA biotechnology and genomics are expected to play an important role in identifying the probiotic strains that are capable of eliciting certain health benefits. rDNA biotechnology also offers the opportunity to decrease or eliminate the allergenic proteins that occur naturally specific foods. For example, rDNA biotechnology has already been used to dramatically reduce the levels of the major rice allergen. Similar approaches could be attempted with more commonly allergenic foods such as peanuts.

MEDICAL BENEFITS

Plants have been a valuable source of pharmaceuticals for centuries. During the past decade, however, intensive research has focused on expanding this source through rDNA biotechnology. The research brings closer to reality the prospect of commercial production in plants of edible vaccines and therapeutics for preventing and treating animal and human diseases. Possibilities include a wide variety of compounds, ranging from vaccine antigens against hepatitis B and Norwalk viruses and Pseudomonas aeruginosa and Staphylococcus aureus to vaccines against cancer and diabetes.

In addition, genetically modified strains of probiotic microorganisms are also possible vehicles for successful delivery of vaccines and digestive aids (e.g., lactase) through the stomach and the small intestine. Human clinical trials suggest that oral vaccination against either of the closely related enterotoxins of Vibrio cholerae and E. coli induces production of antibodies that can neutralize the respective toxins by preventing them from binding to gut cells. Ma et al. showed that tobacco plants could express secretory antibodies or "plantibodies" against the cell surface adhesion protein of S. mutans. Used as a bactericidal mouthwash, the antibodies prevented bacterial colonization by the microorganism and development of dental caries for four months.

A similar approach showed that soybean-produced antibodies protected mice against infection by genital herpes. Compared to antibodies produced in mammalian cell culture, the plantibodies had similar physical properties, remained stable in human reproductive fluids, and exhibited no differences in their affinity for binding and neutralizing herpes simplex virus. Hence, the difference in the glycosylation processes of plants and animals does not appear to affect the immune functions of the plantderived antibodies.

Non-Hodgkins B-cell lymphoma, the most widespread cancer of the lymph system, is difficult to treat because the B-cell tumors are variable and response to treatment can vary from person to person. Hence, effective therapy requires "personalized medicine" tailored to the genetic makeup of each patient's tumor. Unfortunately, conventional treatment methods do not meet the needs for rapid production of customized antibodies in sufficient quantities. Monoclonal antibodies used in conventional treatment also tend to be expensive and unreliable, and those produced in bacteria have solubility and conformation problems.

A system using tobacco mosaic virus (TMV) was developed to produce in tobacco plants a therapeutic vaccine against non-Hodgkin's B-cell lymphoma in a mouse model. Using cells cloned from malignant B-cells of mice, TMV DNA was modified with a tumor-specific sequence from the gene coding for the immunoglobin cell surface marker. Plants were then infected with the modified virus, resulting in expression of cancer-specific antibodies. B-cell proteins were then extracted from the plant leaves for vaccination of the mice. Eighty percent of the mice receiving the plant-derived vaccine survived the lymphoma, while all untreated mice died within three weeks of contracting the disease.

A similar approach was used to develop a vaccine against insulin-dependent diabetes mellitus (IDDM), an auto-immune disease in which insulin-producing cells of the pancreas are destroyed by the cytotoxic T lymphocytes. The "oral tolerance" method of preventing or delaying autoimmune disease symptoms involves the ingestion of large amounts of immunogenic proteins that turn off the autoimmune response. This method of vaccination is gaining recognition as a potential alternative to systemic drug therapy, which is often ineffective.

Insulin and pancreatic glutamic acid decarboxylase (GAD), which are linked to the onset of IDDM, are candidates for use as oral vaccines. The development in a mouse model of a potato-based insulin vaccine that is almost 100 times more powerful than the existing vaccine in preventing IDDM. Feeding diabetes-prone mice potatoes engineered to produce immunogenic GAD reduced the incidence of disease and immune response severity. rDNA biotechnology-derived vaccines are potentially cheap, convenient to distribute, and simple and safe to administer.

Production of medically important substances via rDNA biotechnology engineering of plants and microorganisms offers multiple advantages. For plants, production can be done virtually anywhere and has the potential to address problems associated with delivery of vaccines to people in developing countries. Products from these alternative sources do not require a so-called "cold chain" of refrigerated transport and storage, although they will require segregation from conventional foods to prevent inappropriate consumption.

Pharmaceuticals or therapeutics produced via genetic engineering of plants also offer an alternative delivery method, feeding versus injection, and an alternative to extraction from animal sources. Furthermore, rDNA biotechnology-derived vaccines may also be safer than many conventional vaccines because they consist of pathogen or antibody subunits rather than whole microorganisms. The use of plants can facilitate abundant production of therapeutic proteins without the risk of contamination by animal pathogens, and at substantially reduced cost.

IMPROVEMENTS OF FOOD SAFETY

The potential for food safety benefits from rDNA biotechnology-derived foods and food ingredients. For example, preliminary studies have shown that Bt corn had levels of fumonisin, a potential cancer-causing agent often found at elevated levels in insect-damaged kernels, that were up to 30- to 40-fold lower than in non-Bt corn varieties.

Mycotoxins like fumonisin are both a public health issue and an export issue, as European and Asian markets have refused to import U.S. corn because of what they view as unacceptable levels of mycotoxins. The actual amount of reduction of fumonisin appears to depend on environmental conditions and the specific Bt corn hybrid, but those corn varieties in which the Bt protein is expressed throughout the plant rather than only in specific areas had the lowest fumonisin levels.

Bt corn is modified primarily to resist European corn borers, but it also showed lower mycotoxin levels when corn earworms were present in growing fields. However, the mycotoxin reduction was not as significant as when the primary insect pest was the European corn borer. This preliminary result may lead to the creation of corn varieties with greater resistance to a variety of insects, leading to greater protection from mycotoxins.

Changes to our foods have always produced public concerns. This was the case for hybrid corn, margarine, artificial insemination of farm animals, pasteurization of milk, and microwave cooking, and is the case for rDNA biotechnologyderived foods. The transition from traditional plant breeding to rDNA biotechnology-derived crops has raised several issues that need to be addressed. Upon examination, many of these issues turn out to be without merit.

Economic and access

Public Sector Access: Some critics contend that the increasing role of the private sector in research and aggressive patenting of genes and research resources is limiting the access to the necessary materials and processes for pioneering research in the public sector. A key issue is the scope of patents granted on genes and genetic information as well as on transformation tools and other platform technologies. If concentrated private-sector control of critical genes and technologies becomes a problem, appropriate policy responses include reducing the scope of patents on genes and platform technologies, including obligatory licensing requirements in patent awards, and increasing public-sector funding of basic research to increase the amount of plant genetic information and rDNA biotechnology information in the public domain.

One significant industry response to this concern is one firm's decision to make its extensive rice genome data available to the public for research purposes. In addition, one major biotechnology company recently announced that it will grant patent licenses without charge for the introduction of an rDNA biotechnology- derived crop that will have significant health and nutrition benefits in developing countries. Access to rDNA biotechnology is needed to help meet the need for increasing the world food supply and improving the quality of foods in developing countries.

A strong public-sector agricultural research effort is necessary to provide the benefits of plant rDNA biotechnology to the world's poorest people. The agribusiness input industry will need to find ways to donate technology for use in these poor parts of the world where there are few opportunities for commercial returns. There are organizations that seek to facilitate such transfers, such as the U.S. Agency for International Development (USAID).

Agribusiness Consolidation and Competition: Certain segments of the commercial seed markets have already become highly concentrated. Even in highly concentrated markets, abuses of market power by dominant firms can be restrained by both actual and potential competition. If competing firms can easily enter profitable markets, dominant firms will be prevented from charging exorbitant prices.

There are at least four or five large agricultural and life science companies that are aggressively competing for market share in the corn, soybean, oilseed, and vegetable seed markets. One company has expanded its corn and soybean seed sales in the North American market and has realized strong sales growth for vegetable and horticultural seeds. Yet another company is actively marketing herbicide- tolerant corn and canola seed and has recently established itself as a strong competitor in the vegetable seed market with the acquisition of two smaller companies.

In addition, a new joint venture has been formed for development of advanced cottonseed, and another joint venture is competing in the maize, cotton, and oilseed markets. To its proponents, continued advances in rDNA biotechnology will be needed to feed a growing world population in an environmentally sustainable manner. For example, the further development and application of rDNA biotechnology to agriculture will lead to improved efficiencies in food production.

Strong intellectual property rights are defended as a prerequisite for the private-sector investment needed to realize these potential benefits. Yet some consumer and environmental advocacy groups are concerned that widespread adoption of rDNA biotechnology-derived crop varieties will leave farmers increasingly vulnerable to increases in farm input prices and create unknown risks to the environment and public health. These critics also have concerns about the continuing decline of the small family farmer and the further industrialization of farming. Critics of rDNA biotechnology contend that the combination of strong patent rights for rDNA biotechnologyderived crop varieties as well as technological constraints on seed reproduction could eventually provide major firms with sufficient market power to charge monopoly prices.

But the appropriate public-sector response to potential abuses of market power by agricultural biotechnology companies is not to constrain the use of rDNA biotechnology in agriculture, but rather to maintain a vigilant antitrust policy. To the degree that direct gene transfer may eventually enable crop breeders to incorporate elite traits into even relatively inferior germplasm, advances in rDNA biotechnology breeding techniques could facilitate the entrance of new competitors into profitable seed markets.

However, to promote this type of innovative competition, it will be necessary to maintain widespread access both to the technologies necessary for gene transfer as well as to specific genes that code for important agronomic traits. As a result, the acquiring firm agreed to spin-off its claims to the Agrobacterium method for genetic transformation of corn. As agricultural biotechnology companies continue to increase their investments in genomic research, the requirements for, and scope of, patents granted for genetic information will become an increasingly important policy issue.

Environmental concerns

Environmental concerns have been raised about the impact of pest and disease resistance and herbicide-tolerant plants. All of the new products are carefully tested for safety to mammals and other animal and microbial life. Soil persistence and the likelihood of subsoil water and stream contamination are taken into account by the Environmental Protection Agency (EPA) in deciding whether to register the products for use. Scientists performing these tests and regulators together design testing programmes most appropriate for the new products using the most current scientific knowledge and procedures, as it is very important for all agricultural chemicals to be properly regulated and monitored.

Pest and Disease Resistance: Corn and potato plants have both been successfully transformed with genes from various strains of the soil bacterium B. thuringiensis. These genes encode toxic proteins with specific effects on certain groups of insects. The pollen of Bt plants was to be toxic to the larvae of monarch butterflies feeding on the leaves of milkweed plants. Other studies, however, have shown that in or close to cornfields the concentration of pollen grains found on milkweed plant leaves is, for the most part, well below the threshold level that has any effect on monarch butterfly larval growth or viability.

EPA's recent suggestion that farmers locate the required 20% corn refuge areas around the perimeter of the fields, coupled with the limited movement of corn pollen, would virtually eliminate any remaining risk of Bt pollen to monarch butterfly larvae. Pimentel and Raven assessed the overall picture of the effect on the survival of butterfly populations of Bt corn pollen dusting their larval food plants, and concluded that although Bt corn pollen under certain circumstances has the potential to adversely affect the population levels of Monarch butterflies and other nontarget Lepidoptera, these impacts are minimal compared with habitat loss and the widespread use of pesticides throughout the ecosystem.

The effect of Bt corn on populations of black swallowtail larvae under field conditions, there was no relationship between mortality and proximity to the field or pollen deposition on host plants. They determined that pollen from these same plants failed to cause mortality in the laboratory at the highest pollen dose tested, a level that far exceeded the highest pollen density observed in the field, and concluded that Bt pollen of the variety tested is unlikely to affect wild populations of black swallowtails.

Studies to examine the breakdown of Bt toxin present in debris from corn and other crops with the Bt transgene indicated that the Bt toxin is rapidly broken down by microbial activity and that it has no detectable effects on a range of soil organisms that were tested. However, Crecchio and Stotzky suggested that Bt toxin could persist in the soil bound to humic acids and clay, where it may pose a hazard to non-target insects and enhance the selection of toxin-resistant target species.

A recent rigorous assessment carried out by EPA, however, concluded that plants registered for environmental release that express Bt toxins derived by rDNA biotechnology do not cause unreasonable adverse effects. Since B. thuringiensis is widely used in organic pesticides and is a common soil organism, the exposure of other soil organisms to its toxin is hardly novel.

Resistance to all methods of pest control has been and continues to be a major problem in agriculture.

The levels of expression of the Bt toxin can be 20,000- to 40,000-fold higher via chloroplast gene expression than nuclear rDNA biotechnology-derived plants. These high levels are lethal to resistant insect larvae that can grow on sprayed plants but may be so high as to present other hazards in residues from crop debris.

Transgene Spread by Pollen: There is a concern that the genes for herbicide tolerance may spread via pollen from rDNA biotechnology-derived crops to other native plants. It is theorized that these genes might become established in weed populations, creating forms that would be difficult to control in the future. For soybean, corn, and most other crops in the U.S., this outcome is unlikely because of the absence of related wild species that are either already weeds or have the potential to become weeds.

For those crops that are themselves of weed origin, this problem is a more serious issue. For example, hybrid sugar beet normally takes two years to flower. Its roots are harvested near the end of the first year of growth before flowering. Plants that flower prematurely, or bolt, produce seeds that contaminate the field and give rise to non-hybrid, lower-yielding plants in subsequent crops. In Europe, where bolting is sometimes a problem, a tractor-drawn wick, held above the leaves and soaked with an herbicide, is used to selectively kill the taller- flowering sugar beet plants, which contact the wick. Herbicide-tolerant bolters will reduce the choice of herbicides that can be used in this way. In some rice-growing regions, red rice is a weed in rice paddies.

Because it readily hybridizes with cultivated rice, it would be very unwise to use herbicide-tolerant rice in such regions, since the red rice population would rapidly acquire herbicide tolerance, denying rice farmers a tool for controlling it. It is also generally regarded as unwise to produce herbicide-tolerant sorghum for use in the U.S. because of the likelihood of outcrossing to Johnson grass, which is a particularly difficult weed to control in agriculture.

In Canada, volunteer canola plants that arise from spilled seed from the previous season's crop can be a problem for subsequent crops of canola and other species. Canola resistant to three commonly used herbicides have arisen in Alberta, Canada, from intercrossing among two adjacent crops that gave rise to doubly resistant volunteers that, in the following year, crossed with a nearby crop tolerant to the third herbicide.

Farmers usually seed a cereal crop after canola and apply a phenoxy herbicide to kill canola volunteers. If the treatment is done properly, the herbicide-resistant volunteers will be eliminated. Although the triply resistant forms can be controlled by other herbicides, such as 2-4 D, their origin points to a failure to manage the release of the original varieties in a way that will safeguard their continued usefulness in agriculture. In short, growing adjacent crops with different transgenes that are better kept apart is not prudent. A relatively new technology introduces transgenes, such as those for herbicide tolerance, in the DNA carried by plant chloroplasts.

Chloroplasts usually are not carried in pollen grains, so the spread of these transgenes is limited to the seeds produced by the rDNA biotechnology-derived plant. This technology shows promise in restricting the spread of transgenes. Chloroplast transformation is technically more difficult to carry out than nuclear transformation, so rDNA biotechnology-derived crops produced in this way are not yet commercially available. Herbicide tolerance that has spread to weeds is very unlikely to be a problem in the absence of selective pressure from herbicide application and is thus unlikely to be a threat that extends beyond agriculture and cultivation.

There is some concern that transfer of genes like Bt to related wild species by out-crossing could increase their competitiveness and lead either to enhanced weediness or to undesirable changes in the wild population. In parts of Mexico, corn and teosinte freely intercross in farmers' fields. If insect damage exercises significant control of the teosinte population, there could be strong selection for resistance with undesirable consequences for those farmers. However, conventional breeding for insect resistance in corn has not engendered similar concerns.

Organic Farming: The organic farming community has decided at this time not to use rDNA biotechnology-derived crops. Thus, if an organic crop, grown for its harvested seed, is planted near an rDNA biotechnology-derived crop of the same species, it is likely that some seeds will result from fertilization by pollen carrying a nuclear transgene. With the sensitive DNA detection techniques that are now available, if the transgene signature is detected, it could invalidate the crop's organic certification. Chloroplast-encoded transgenes will avoid this problem, but so will observing reasonable isolation distances between crops. These isolation distances are used to ensure the genetic purity of named conventional crop varieties grown for seed.

Virus Resistance: The ability to confer viral resistance by using transgenes that incorporate a part of the viral genome, such as a gene encoding the viral coat protein, or a gene responsible for organizing the movement of virus particles from cell to cell through the minute pores that connect them, has had a dramatic impact on several crops. For example, in Hawaii the papaya industry was devastated by ring-spot virus, which has now been successfully controlled by planting rDNA biotechnology-derived papaya with a gene encoding the viral coat protein (Gonsalves, 1998). The viral coat protein gene is used because, above a certain concentration, in or on the plant, the viral coat protein inhibits further growth of the virus. Presumably, this trait evolved so that the virus did not kill its host too promptly.

In the U.S., squash plants resistant to the aphid-borne zucchini yellows virus have been developed and provide effective control. This example has been criticized on the grounds that tests which found that native populations of wild squash relatives (cucurbits) did not harbor the virus and therefore are unlikely to be controlled by it, were on too small a scale. Others have pointed to examples where multipartite viruses may be reassembled by crossing different rDNA biotechnologyderived parents that carry the separate components. It can be theorized that recombination in an rDNA biotechnology-derived host plant between a systemically expressed viral component and the genome of another, randomly infecting virus might result in a new form that could create a serious problem.

While conceivable, opportunities of this kind occur all the time when plants become naturally infected with more than one kind of virus. Recombination between a virally derived transgene and another virus has been suggested as a possible source of a new virus with enhanced virulence. Such recombination has been shown in laboratory studies, especially with high selection pressure. A recent NRC committee concluded that "most virus derived resistance genes are unlikely to present unusual or unmanageable problems that differ from those associated with traditional plant breeding for resistance".

The S35 promoter from cauliflower mosaic virus (CaMV) is used in almost all commercialized rDNA biotechnology-derived crops. The CaMV35S promoter will cause large-scale genomic rearrangement, with the extreme suggestion that it could cause cancer. Scientists knowledge able about CaMV35S note that about 10% of the cauliflower and cabbage produced are infected with CaMV, thereby providing 10,000 times greater 35S promoter in the diet than in rDNA biotechnology-derived crops. No evidence of 35S promoter transfer has been observed, in spite of human consumption of CaMV35 in cauliflower and cabbage throughout history.

Monoculture and rDNA Biotechnology: Conventional plant breeding and improvements in agronomy have helped farmers to maximize yields and profits. Yield trials of the crop varieties available usually reveal one-or a few that are very similar-that is best for an agricultural region. The dense, uniform crop stands that cover very large areas are an invitation to epidemic pests and diseases that are kept in check by breeding for resistance and by pesticide applications. The Southern corn leaf blight epidemic of 1969-70 revealed the inherent weakness of a crop whose hybrid seed production depended on cytoplasmic male sterility to avoid removing the pollen-bearing tassels.

A similar level of dependence on a particular transgene could easily arise, as has been shown by the very rapid adoption in the U.S. of herbicide-tolerant soybean and insect-resistant corn. Even though the risk is probably not high, it would be prudent to adopt the same strategy advocated following the Southern corn leaf blight, which was to diversify the germplasm. This is to make sure that there is adequate backup capability that can provide alternative varieties in the event of catastrophic failure. The best way of doing this is still to safeguard germplasm collections and to encourage a broad spectrum of plant breeding activities.

Monitoring concerns

Food Safety Monitoring: The initial safety evaluation of rDNA biotechnology-derived foods addresses both short-term and long-term potential food safety issues. The issue of longterm human food safety was considered by a recent consultation of the Food and Agriculture Organization of the United Nations (FAO) and the World Health Organization (WHO). The consultation concluded that the possibility of longterm effects being specifically attributable to genetically modified foods is highly unlikely.

In considering the issue, the consultation noted that very little is known about the potential of long term effects of any foods. In many cases, this is further confounded by wide genetic variability in the population, such that some individuals may have a greater disposition to food-related effects. The consultation acknowledged that for genetically modified foods, the premarketing safety assessment already gives assurance that the food is as safe as its conventional counterpart. Furthermore, it was recognized that observational epidemiological studies would be unlikely to identify any such effects against the background of undesirable effects of conventional foods.

Environmental Monitoring: Genetic modification of plants through plant breeding is well established as a major contributor to increased yield and value addition to our food and feed supply. Plant breeding is conducted in both the public and private sector, with the private sector in recent years dominating the major crops, e.g., corn, cotton, sorghum, and soybeans. Extensive field testing involving multiple sites and several years typically occurs before seed multiplication and release of a commercial seed product. Most states test and release public varieties in a tracked process that produces certified seed.

Most states also provide public testing in a fee-based evaluation at a few sites. Private-sector seeds are sometimes submitted for public testing, but usually undergo similar field evaluations and, in the case of large seed companies, may be tested at hundreds of sites. The ultimate test in all cases is in commercial farmers' fields. Almost all farmers monitor their crops from planting to harvest by regular field examinations. This key monitoring activity is done by the growers or, in the case of large acreage, may be performed by hired consultants or scouts.

Consultants or scouts are frequently used for cotton but are less common for soybeans. This "on-the-farm" monitoring follows emergence, growth, nutrient limitations, flowering, fruit set, maturation, insect pests, disease, weed control, and other key events. Finally, the yield for each crop in each field is measured at harvest.

The experience of the farmer or consultant for specific crops in specific locations represents a baseline. In addition to the farmers and consultants, there are also extension agents and industry representatives who do a modest amount of monitoring. Monitoring by the farmer or hired consultant has proven to be effective and serves as the first alert for untoward effects. For example, farmers were the first to identify herbicide-tolerant weeds, e.g., atrazine-resistant pigweed in Ontario in the 1950s to 1960s. Untoward environmental or performance effects of rDNA biotechnology-derived seeds-as is the case for traditional seeds-would normally be identified by the farmer or consultant.

Examples are the reduced boll set of herbicide-tolerant cotton in the Mississippi delta in the introductory year of this product, as well as periodic herbicide performance problems. Special monitoring by the seed company is required for Bt crops to provide early identification of Bt-resistant insect pests. The farmer and consultant monitors, however, are likely to provide an early alert for actual pest resistance.

The road map for rDNA biotechnology- derived seeds from initial generation through testing and regulatory approval is similar to that for traditional seeds, with additional steps to meet regulatory requirements. To date, almost all the commercialized rDNA biotechnology- derived seeds have been developed by the private sector, with four large firms as the primary technology developers. From several hundred up to a thousand transformants with the desired gene or trait, one or a few are selected for development. The selected transformants are evaluated for efficacy in standard and other genetic backgrounds, agronomic characteristics, and genetics. Equivalency testing is performed at multiple locations by the seed company.

In addition, the rDNA biotechnology- derived seed is analyzed to pro- vide data for regulatory review. Questions remain as to long-term effects on organisms (e.g., birds, plants, animals) and microorganisms in the environment. These questions are the same as those that need to be raised for introducing any new types of plants or even new varieties of an established plant. Hypothetical deleterious changes in producing modified foods are possible, as such changes have occasionally occurred in nature and in conventional plant breeding. Such plants rarely make it to the marketplace and, if they do, can be readily removed.

With a higher degree of regulatory oversight for all foods derived by rDNA biotechnology, there is less likelihood of adverse reactions to consumers than with conventional foods. rDNA biotechnology will require continued research and management, as well as monitoring and surveillance, to produce high-quality and affordable foods.

Food allergies

Food allergies involve abnormal immunological responses to substances in foods, usually naturally occurring proteins. The majority of food allergies are traced to eight commonly allergenic foods or food groups: milk, eggs, fish, crustacea, peanuts, soybeans, tree nuts, and wheat, although other sources of genetic material can possess genes encoding for environmental allergens such as pollen allergens. Allergic reactions can be manifested by symptoms ranging from mild cutaneous or gastrointestinal symptoms to life-threatening anaphylactic shock reactions.

Virtually all food allergens are proteins, although only a small fraction of the proteins found in nature (and in foods) are allergenic. Since genetic modifications involve the introduction of new genes into the recipient plant and since these genes would produce new proteins in the improved variety, the potential allergenicity of foods developed through rDNA biotechnology has been a source of some concern.

Despite the concerns, no unique allergic reactions have yet occurred to any of the foods derived through rDNA biotechnology. Of course, a consumer with a soybean allergy is likely to be reactive to an rDNA biotechnology-derived soybean as well. But no new and novel allergens have been introduced into foods through rDNA biotechnology. In fact, the proteins introduced into rDNA biotechnology-derived foods to confer traits such as insect resistance and herbicide tolerance are unlikely to be allergenic because they are expressed at very low levels in the modified food, they have no amino acid sequence homology to known allergens, and they are readily digested.

The potential allergenicity of rDNA biotechnology-derived foods can be assessed using a decision-tree strategy developed by the International Food Biotechnology Council (IFBC) and the Allergy and Immunology Institute of the International Life Sciences Institute (ILSI) in 1996. The utility of this approach was recently recognized by FAO/WHO.

This strategy focuses on specific scientific criteria, including the source of the gene(s), the sequence homology of the newly introduced protein(s) to known allergens, the immunochemical reactivity of the newly introduced protein(s) with immunoglobulin E (IgE) antibodies from the blood serum of individuals with known allergies to the source from which the genetic material was obtained, and the physicochemical properties (e.g., digestive stability) of the introduced protein.

If genes are obtained from known allergenic sources, the possibility of the transfer of a known allergen must be carefully examined. The potential hazards are illustrated by the case of a soybean variety constructed to correct the inherent methionine deficiency existing in soybeans. A high-methionine protein was introduced into soybeans by one firm using a gene from Brazil nuts. Brazil nuts are known to be allergenic, but, at the time of this development, the allergens from Brazil nuts had not been identified. The high-methionine protein from Brazil nuts was identified as the major allergen in research sponsored by that firm. As a result, commercial development of that particular soybean variety ceased.

Genetic transformation

Genetic transformation of plant cells is an inherently infrequent event. The challenge that the researcher faces is to identify the few cells that have integrated the introduced DNA from a large population of non-transgenic cells. This is most often done by introducing a selectable marker that permits growth of only cells containing the newly introduced DNA. In plant transformation, a gene for resistance to the antibiotic kanamycin dominated early rDNA biotechnology- derived crops.

Concerns have been raised about the potential for horizontal gene transfer of the antibiotic resistance gene from an rDNA biotechnology-derived plant to microorganisms, thereby reducing the efficacy of the antibiotic. However, both scientists and most regulators around the world generally believe that this risk is virtually nonexistent. The conclusion derives from a number of facts. First, the marker gene has been altered to express in plant cells.

Genes are not unique to specific organisms, the controlling elements that permit gene expression are very different in plants and microorganisms. One would not expect that a gene engineered to work optimally in plant cells would work effectively in bacteria. Second, the antibiotic-resistance genes are stable when integrated into plant DNA. Plant DNA, upon exposure to the gastrointestinal environment, would be rapidly hydrolyzed to small, nonfunctional pieces long before it came into contact with microflora.

Third, DNA uptake into bacteria is an extremely inefficient process requiring either transformation competence or specific DNA transfer mechanisms employed between bacteria. There are no known mechanisms for transfer of DNA from plant cells to bacteria, and the bacteria in the digestive system would not be competent to take up free DNA. Fourth, even if such mechanisms for DNA uptake were in place, stable integration of that DNA into bacteria requires extensive DNA sequence homology between the incoming DNA and the host chromosome.

Such homology would not exist unless bacteria already possessed the antibiotic resistance gene prior to DNA uptake.

Fifth, even if such unlikely transfer were to occur, positive selection pressure would be required; e.g., the person would have to be taking the antibiotic to which the resistance was encoded at the time of such transfer. Finally, there are no authenticated reports of any horizontal DNA transfer occurring from food plants to bacteria within the gastrointestinal tract of humans.

Even if this occurred by some unknown mechanism at some vanishingly small frequency, there would be no consequence, because of the existing level of antibiotic resistance already present in gut microflora. A recent FAO/WHO joint consultation addressed the concern that there might be transfer of antibiotic resistance from the widely used antibiotic resistance marker genes and concluded that no health risk is presented:

The Consultation considered horizontal gene transfer from plants and plant products consumed as food to gut microorganisms or human cells as a rare possibility, but noted that it cannot be completely discounted. The most important consideration with respect to horizontal gene transfer is the consequence of a gene being transferred and expressed in transformed cells.

The Consultation further noted that the antibiotic resistance markers currently used in genetically modified plants have been previously reviewed for safety. It has concluded that there is no evidence that the markers currently in use pose a health risk to humans or domestic animals. In addition, non-antibiotic resistance markers have mainly replaced kanamycin in products now in the pipeline. These include removable selectable marker genes such as using the Cre-lox site-specific recombination system or transposable elements. Cre is a recombinase; lox is a 32-base-pair recognition site. Positive selection systems will probably dominate in the future.

One system uses an exclusive energy source, cellobiuronic acid, a disaccharide that, when transported into the cell, is metabolized to glucose by beta-glucuronidase.

Naturally Occurring Toxic Substances

The great majority of food plants, and many animals used for food, produce or carry naturally occurring toxic substances. The only categories of organisms used as human food that have essentially no or only a very rare content of naturally occurring toxicants are the cereal grains and domestic animals. The absence of toxicants from these food sources is entirely due to man's interference with nature-millennia of selective breeding and centuries of careful husbandry have reduced their original toxicant content.

Plants,. and many animals, produce toxicants for a variety of reasons. Some kill or repel predators, pests, or diseases. Others are pollenator attractants. Some inhibit competitive species. Others are metabolic "dead ends"- a means for a plant to sequester a plant toxicant it can neither avoid nor excrete. The largest known number of naturally occurring toxicants are endogenous, or "constitutive"; i.e., they are produced by the normal metabolic processes of the organism that is the food source.

An example is solanine, a neurotoxin in potatoes that has been the cause of numerous outbreaks of human poisoning when potatoes were grown under unfavorable conditions or when they formed a large part of the diet. Another example is cyanogenic glycosides, found in several foods such as lima beans and bamboo shoots. In these and other crops, conventional breeding has been used to decrease toxicant levels.

CHAPTER 5

SAFETY OF GENETICALLY MODIFIED FOODS

Food produced from genetically modified (GM) crops has become one of the biggest disputes in the already controversial area of food and agricultural production. The level of concern and resistance to GM foods from consumers appears to have shocked the biotechnology industry and many scientists, yet it is clear from this early public opinion research that concerns about the safety of GM foods are deep seated. They range from the moral and ethical implications of genetic modification, to concerns about the environmental impact of the spread of modified genes, from changes in agricultural practice to the impact on poor farmers in developing countries. They include fears about the safety of the foods derived from GM crops.

The scientific and political establishments have been consistently slow to accept the public's concerns and attempts to address them have been inadequate. Consequently, claims that regulatory procedures ensure the safety of GM crops have remained unconvincing. At least in the UK, the credibility of Government and industry has been severely weakened by the mishandling of BSE in UK cattle and the subsequent development of new variant Creutzfeld Jacob Disease in humans.

CHALLENGE OF GM CROPS

For thousands of years, farmers and plant breeders have been changing crop plants to improve characteristics such as size, appearance and taste. This was done by taking advantage of the natural variation found in the different populations of crop plant species. Plants which grew well, had a higher yield or tasted better, were selected and bred from. This process, known as selection breeding, is still the most widely used technique for developing new plant varieties within a particular crop species. Between different species there are natural barriers which prevent cross breeding, so selection breeding is only able to make use of the genes and characteristics that are naturally present in the different populations of one crop species.

Over the centuries, these valued characteristics have been concentrated into particular varieties by the breeding process. By the middle of the twentieth century, plant breeders were seeking ways to extend the range of characteristics available to them. For example, attempts have been made to induce mutations in the genetic make-up of plants, by irradiation or the use of chemicals, in the hope that this would create new and useful characteristics in the plants. But the use of such techniques has not been widely adopted, mainly because most mutations cause worse rather than better characteristics in the plants, and by 1991 'mutation breeding' had only produced 1,500 registered varieties of plants and micro-organisms around the world. In their most recent attempt to improve our food crops, scientists and plant breeders embarked on the process of genetic modification, in which genes are moved between unrelated organisms, with the aim of introducing specific new traits of commercial value. The differences between this and selection breeding are explained below.

Getting Novel Genes into Plants

Within every cell of every organism is the biological code for its construction. This code is held within the structure of an extremely long chain molecule, deoxyribonucleic acid (DNA), which contains all the information necessary for the growth, maintenance and development of the organism. This code can be broken down into small segments controlling specific functions, and these are referred to as genes. Very simple organisms, such as bacteria, may have a few thousand genes, while more complicated organisms have many more - it has been estimated that maize has around 50,000 genes.

The long DNA molecule is tightly packed into a structure called a chromosome, and higher organisms have more than one of these tightly packed bundles of DNA, each one containing masses of genetic information. The collective term for all of the genetic information in an organism is the genome. Transferring DNA and genes from one organism to another is a difficult procedure. In early attempts it was realised that one of the main problems would be getting past the natural defences of plant cells.

Different techniques were developed in order to do this, such as shooting tiny metal balls coated with DNA into cells (particle bombardment) or mixing DNA with cells from which the cell walls had been removed (protoplast transformation). But the most commonly used technique involves a bacterium called Agrobacterium tumefaciens (A tumefaciens). This bacterium already possesses the ability to get past plant cell defences, as it causes cancerous growths in plants, called crown galls. It does this by transferring its own genes into plant cells through wounds and the genes incorporated into the chromosomes of the plant cells, causing them to become cancerous.

Genetic modification effectively hi-jacks A tumefaciens: in order to get novel genes into plants, the cancer causing genes are removed from A tumefaciens, and novel genes put in their place. The bacterium then transfers these into the plant cell. After this transfer has occurred, the bacteria are killed using an antibiotic. However, recent research for the UK Government found that A tumefaciens bacteria are not killed by antibiotics that have been applied after the genetic modification of plant cells has taken place. Regular checking for the presence of the modified agrobacteria does not take place during the development of GM plants - it is assumed they have been killed. Yet it was found that the bacteria could survive for up to 13 months inside the transformed plants, as well as in plants derived from them, and after six months 12. 5 per cent of these bacteria still contained the novel genes. They were found to be still capable of transforming cells in the plants they were infecting.

Haphazard Process

Whatever means is used, the insertion of novel genes into the plant cell is a fairly haphazard process. There is no way to control or direct what happens - new genes end up being inserted randomly into the genetic makeup of the host organism. In fact, the operation to insert new genes frequently fails - inserted genes fail to work, behave in ways that aren't expected, or the functioning of native genes may be affected. There is no means of controlling where the genes will end up within the chromosome, and in many cases more than one copy of the gene may be incorporated, or extra DNA from the bacterial plasmid may get incorporated as well as the novel genes.

In many organisms, including wheat, maize and rice, DNA is divided into active regions of working genes and the so-called junk DNA that does not show any activity. In a recent experiment using rice, it was found that novel genes only integrated into this active, gene rich region of the genome. In other words, the novel genes always ended up in the area where there were already genes working. It is now known that genes do not work in isolation, in fact they often work in groups, and their neighbouring genes can affect how they work. Because of this, the insertion of a novel gene into an active genetic region will always have the potential to cause disruption to the genetic functioning of the organism.

Many Crops - the Same Genes

It is only rarely that genes are sourced from the crop species being modified, and often they come from extremely distant organisms, such as bacteria. Sourcing and developing novel genes can be a time-consuming and expensive process so, once isolated, the same genetic sequences are used time and again. This means that the novel proteins produced by these genes end up in a wide range of GM crops.

Under natural conditions, the characteristics of a crop plant, such as the size of fruit, are controlled by groups of functionally related genes, which provide a co-ordinated control of the characteristic. These genes are bound up with associated 'regulatory' sequences of DNA, which ensure that the genes work in the right part of the organism and at the right stage of its life cycle. These have evolved along with the genes they are attached to and so they are specific to the plant species, or even to the genes, providing a very tight control over the functioning of the genes in the organism. During traditional selection breeding, these groups of genes are moved around within the plant population and concentrated into a particular plant line by crossing different varieties together.

In contrast, genetic modification tends to use single genes from species other than the crop plant to get the desired characteristic. These novel genes also need regulatory sequences attached to them so that they will work in the plant. But very few of the regulatory sequences in plants have been fully described, and without the right regulatory sequence the introduced genes will not function as desired. To get around this problem, genetic engineers have been forced to use short cuts - such as using regulatory sequences from viruses. Viruses attack organisms by taking over the genetic machinery of an infected cell, and reprogramming it to make more viruses.

To achieve this, the virus genes must have regulatory sections that the genetic machinery of the infected cells will be able to understand. This has proved convenient for genetic engineers, and viral regulatory sequences are often attached to the novel genes inserted into GM crop plants. The viral regulatory sequence allows the novel gene to work in the GM plant even though it hasn't got the 'right' regulatory sequence from the plant itself.

One of these viral regulatory sequences is known as the 35S promoter sequence, taken from the plant pathogen Cauliflower Mosaic Virus (CaMV). It has proved to be very successful at allowing inserted genes to function, but because it comes from a virus rather than the crop plants, it is not nearly as precise as the native regulatory sequences. In fact, the CaMV promoter is extremely powerful and, if inserted successfully, causes the inserted novel genes to be 'on' in every cell of the plant, as well as allowing expression of the novel gene at very high levels. It is effective, but it is a blunt instrument in terms of genetic regulation.

In contrast to the precise and controlled action of the native genes, inserted genes linked to the CaMV promoter may be on at full blast in every cell of the organism from the moment of fertilisation through to death - meaning that the novel protein is produced at high levels in the plant through out its life. This is completely at odds with natural genetic functioning and the consequences are largely unknown. More precise regulatory sequences are being developed, but CaMV is still the most commonly used promoter in GM crops, in part because it is in the public domain and therefore there is no need to pay royalties for its use. In fact, the popularity of this blunt instrument promoter is illustrated by the fact that the vast majority of GM crops on the market, or close to marketing, contain this construct.

Because novel genes are inserted at random, they can have unpredictable and unexpected impacts upon the genetic function of the host organism. There is increasing evidence that genes interact and work with each other rather than being single isolated units and the inserted novel genes can impact on the functioning of nearby native genes. Alternatively, native genes can affect the functioning of the inserted genes, and in some cases cellular defence mechanisms may cause inserted genes to be switched off - referred to as silencing.

At present, the function of most native genes in crop plants is unknown - gene mapping can establish what genes are present, but even then the function of mapped genes is often unclear and scienitific knowledge is a long way from explaining which genes are working together and how they interact. There is no guarantee that inserted novel genes will behave as expected in every case. Concern has been widely expressed that their insertion could disrupt the genetic functioning of GM crop plants in such a way that unexpected toxins could be produced, or the levels of nutrients and naturally occurring toxins altered. Theoretically, the possibilities include the disruption or alteration of the expression of genes native to the plant - increasing or decreasing levels of naturally occurring nutrients or toxins.

The novel gene could also cause a naturally occurring but silenced gene to be switched on, perhaps producing an unexpected toxin or new biochemical pathway within the plant. Or, production of a novel protein as the result of gene insertion could divert resources away from the normal metabolic pathways of the plant. This is not only a theoretical concern - there are many examples where genetic engineering has altered the composition of GM plants in entirely unexpected ways.

EFFECTS OF NOVEL GENES

Molecular analysis techniques, such as 'southern blot' analysis, are used to detect the presence of the inserted novel gene in the genetically modified organism. They can also be used to give an indication of where the novel gene is located in the genome of the host organism. Where the inserted novel gene has ended up determines whether or not it will have a disrupting effect on native genes in the plant and so, While gene maps for the major crops, such as rice and maize, are close to completion, understanding of what the genes do and how they interact will take many more years to establish.

In the absence of such information the real impact of putting new genes into an organism cannot be known. Knowledge is still too limited to be able to establish whether or not there may be side effects, then perhaps we should consider whether the whole push for commercial GM crops is premature. In addition, it is becoming apparent that the techniques used for analysing the GM crops already on the market did not provide full or accurate information about inserted genes. More sensitive techniques need to be developed and applied to the molecular analysis of GM crops and foods. It is essential that the first step in the safety assessment of GM organisms should be to establish whether or not the inserted novel gene is having an effect upon the host's genetic functioning.

Marker Genes

Whatever means is used to genetically modify a plant cell, the success rate is very low and it is necessary to have a means of identifying those cells which have been successfully modified from the rest of the cells which have not. This is done by use of a marker gene, which is attached to the novel gene, and which confers a characteristic that is easily tested for. The addition of a gene for the production of â glucuronidase (which produces a blue colour), for example, would allow the transformed cells to be selected visually.

However, one of the most common techniques so far has been the use of antibiotic resistance genes. In this case, the transformed cells gain resistance to a specific antibiotic, so that when it is applied the transformed cells survive while all the others are killed. However, these marker genes are usually retained in the plants after they are developed for use as a food crop, and this has been a source of continuing controversy.

Antibiotic Resistance Markers (ARMs)

The biggest concern about antibiotic resistance markers (ARMs) is the possibility that they could be transferred to bacteria in the guts of animals or humans, or to bacteria in the environment. Many bacteria have the ability to pick up genes from their surroundings and to pass these genes on to other species of bacteria, including those which cause disease. There is a danger therefore that diseases could become resistant to important antibiotic drugs.

The use of antibiotic resistance genes has also been criticised by the Royal Society, the UK Government's Advisory Committee on Novel Foods and Processesand the National Farmers Union. In contrast, the US FDA has remained unruffled about the presence of antibiotic resistance markers in GM foods and, by the time it produced guidance on the subject in 1997, it had already approved 19 GM crops containing one or more antibiotic resistance marker genes. Marker genes giving resistance to the same antibiotics appear in a wide range of crops. For example, the nptII gene gives resistance to the antibiotics neomycin, gentamycin B and kanamycin. It is argued that the use of this marker gene is not worthy of concern because many bacteria in the environment already have resistance to these antibiotics.

The biggest concern lies with ampR marker gene, which confers resistance to antibiotics in the ampicillin family. This gene is found in Novartis' insect-tolerant Bt 176 maize and Aventis' T25 herbicide-tolerant maize (although only partial copy of the gene is present) which are approved for use in food and animal feed in the EU and the US. The gene is also in Aventis' T14 herbicide-tolerant maize, Dekalb's DTB418 maize and Dupont's high oleic acid soybean which are only approved in the US. Novartis' Bt176 maize is grown in Spain and Portugal.

In bacteria, the origin of replication controls how many copies of the gene they can produce - the more copies of the gene there are, the more protein is produced and so the greater the power of the bacteria to resist the ampicillin antibiotics. The origin of replication in Novartis' maize was artificially constructed and is not found in nature. It would allow bacteria to produce more than 600 copies of the antibiotic resistance gene. The naturally occurring version would only allow them to produce between four and 18 copies. This means that if bacteria acquired these genetic constructs from the Novartis maize, they would become resistant to the antibiotics in the ampicillin family.

Most GM food for humans is highly processed. Processing tends to break down the DNA in food, making it less likely that bacteria would be able to pick up ARMs from GM food. However, animal feed is much less processed and this may be a route for the transfer of GM antibiotic resistance genes into bacteria in the food chain. Whilst this caused concern to the UK Government's scientific advisors, it appears to have been taken far less seriously by the companies producing these crops. For example, maize is commonly made into silage by being partially fermented in a store and then fed to cattle.

In 1996, Sharpes International Seeds made an application to the UK Government to release Aventis' T14 GM maize, (which contains the ampR gene). Researchers in the UK have found that bacteria in the mouth are capable of taking up and expressing naked DNA from saliva,so it is possible they could incorporate genes from GM food when it is eaten. In the Netherlands, researchers used a model of a human gut to look at what would happen to GM food after it is eaten. They predicted that six per cent of the genes from GM tomatoes would survive digestionand considered that the genes could survive for long enough for bacteria to pick them up. It has even been suggested that bacteria in the respiratory tracts of workers could pick up DNA from pollen or flour during production and processing of GM crops. Although the relative frequency of such events is likely to be low, the ACNFP has pointed out that "given the huge amplification of resistance genes implicit in the agricultural application of transgene technology, even rare events will happen".

SUBSTANTIAL EQUIVALENCE

Over the last ten years there has been a succession of serious incidents affecting public health that were brought on by changes in food production. Examples include the rise of salmonella infection in UK poultry flocks to epidemic levels, the rise in E coli 0157 poisoning around the world, and of course bovine spongiform encephalopathy (BSE) in UK cattle and the subsequent development of new variant Creutzfeld Jacob Disease in humans. All these incidents have made it increasingly clear to consumers that, in a world where food production operates on large scales and is widely industrialised, small changes can lead to disastrous consequences.

Researchers in the UK found that consumers viewed food safety issues as longterm and complex, and that on such issues "studies would have to be undertaken for many years, maybe even a generation, before people would be convinced". And, while the biotech industry has consistently claimed that GM foods are as safe as any other, government regulators have put in place procedures for examining the safety of GM crops and foods. In recent years, a heated debate has developed about their effectiveness and appropriateness. In particular, attention has focused on the concept of substantial equivalence, upon which much of the assessment procedure is based.

Since the early 1990s, various international industry and governmental organisations have worked to design strategies for evaluating the safety of GM foods, including the International Food Biotechnology Council (IFBC), the Organisation for Economic Cooperation and Development (OECD), the United Nations Food and Agriculture Organisation (FAO)and the World Health Organisation (WHO). Although these strategies have not had legal status, the principles evolved have been incorporated in varying forms into national legislation.

The common methodology for these different strategies and national legal requirements has been a comparison between the GM food and a non-GM counterpart. The underlying rationale for this comparative assessment (rather than a safety assessment of the GM food in isolation), is the assumption that existing foods have a long history of safe use and so, if a GM crop is found to be the same as the non-GM counterpart, it too should be able to lay claim to this history of safe use. This approach, referred to as substantial equivalence, has come to underpin the safety assessments of GM foods around the world.

Substantial equivalence was developed as a response to the difficulties encountered when standard procedures, used for testing the safety of single chemicals, were applied to foods. All foods are made up of a complex mixture of different compounds, which provide their energy and nutritional value. They contain beneficial nutrients, such as protein and starch, which provide the major nutritional value, as well as vitamins and minerals. But they may also contain compounds which are not beneficial, ranging from the caffeine in tea and coffee to the poisonous glycoalkaloids present in green potatoes.

Foods may also contain anti-nutritional compounds, such as phytate in wheat which prevents the absorption of iron, or they may contain allergens like those in peanuts. It is the combination of all these components which determines the safety and nutritional value of a food and its products, and genetic modification has the potential to affect any of these. This complexity made the existing toxicological techniques used for single compounds extremely difficult to apply. Procedures for testing the safety of new chemicals and medicines have been in use for decades.

However, these were developed to examine the safety of a single compound, not for testing a range of compounds which will all be eaten at once, as is the case for GM foods. It is extremely difficult to isolate the effect of any one compound in a food, and impossible using the techniques available at the beginning of the 1990s to look at the effects of all the different compounds in GM food at the same time. For example, a fairly crude safety test for any compound is the high dose feeding trial.

In the test, rats or mice are fed higher and higher doses of a new drug, food additive or chemical, until a toxic effect is observed. This is supposed to provide a starting point for the examination of the nature and level of any toxic effects. But even this crude examination cannot be applied to GM foods. For example, in one study freeze-dried GM tomatoes were fed to rats, at a rate equivalent to 13 fresh tomatoes per rat every day. But even this incredibly high dose was still not enough to satisfy the dosage requirements normal for a single compound. It was simply not possible for the rats to eat enough tomatoes and, even if they could have done so, their diet would have been so unbalanced that this alone would have made them ill.

While the methods used for food additives and medicines turned out to be largely inapplicable to whole foods, historically there were few attempts to assess the safety of conventional foods. This meant that there was no experience upon which to draw and a huge gap was found between the ability to create GM foods and the ability to test their safety. This was recognised by the scientists at the OECD and the WHO was charged with coming up with procedures for assessing GM food. But instead of taking the clear, strong position of pointing out that there was no adequate way to assess the safety of GM foods, the WHO compromised and came up with substantial equivalence.

The safety of non-GM foods is based largely upon their history. It is generally accepted that if there is a history of safe use for a food, such as rice, wheat or potatoes, then the food is safe for most of the population. The scientists reasoned that if a GM food could be shown to be the same as a conventional food, it would also be able to lay claim to this history of safe use. The problem with this was that GM foods, most of which contain compounds that have never been eaten before, are never entirely the same as a conventional food. To solve this contradiction, it was decided that only a degree of 'sameness' would be required.

What degree of sameness would suffice remained uncertain and so it was decided GM food would be allowed to lay claim to the safety of a normal food, if it was shown to be mostly the same as its conventional counterpart - if it was substantially equivalent to it. In 1994, the OECD developed the methodology that underpins GM food safety assessment in the industrialised nations,and it based this on the concept of substantial equivalence. It established the principle that the greater the equivalence of a GM food to its conventional counterpart, the smaller the requirement for testing.

Foods in Future

The next generation of GM foods will have no equivalent. For example, take the case of the recently developed beta carotene-producing rice. How will this be compared to non-GM rice, which does not share the metabolic pathway that produces beta carotene? The greater the difference between the GM crop and its non-GM counterpart, the more difficult the concept of substantial equivalence becomes. The EU Scientific Committee for Food attempted to tackle this problem by suggesting that the assessment "contains a dynamic element, as the continuing modification of a food requires that the basis of comparison will evolve in a way that the most recent NF [Novel Food] is compared with an appropriate former NF and not necessarily with the most traditional counterpart".

In other words, as GM crops and foods become more different from conventional ones, their substantial equivalence should be assessed against existing GM crops and foods. This proposal does little to solve the inherent difficulties of substantial equivalence, but does raise the possibilities that any mistakes made with one generation of a GM crop could be compounded in the next. The idea that a new GM food would be deemed to be acceptable because it is substantially equivalent to a GM food that was substantially equivalent to a non-GM food does not provide much comfort.

The issue of substantial equivalence and how it affects food safety is of even greater importance for developing countries. Rich people's diets are usually varied - it is unlikely that one single food will make up a large percentage of the diet. But for the poor, diets are restricted by what is affordable, and a single food can make up an extremely large proportion of the diet. A report by the UN FAO found that maize provides up to 45 per cent of the daily calories and 59 per cent of the daily protein for both rural and urban populationsin Central America. It is widely proposed that GM crops be introduced in developing countries to feed the poor. This would mean that GM foods would make up an excessively large proportion of the food of the poor.

It is comments made by the US Government with respect to their animals that are probably of most relevance in this case. The Director of the Centre for Veterinary Medicine stated that GM crops would be of particular concern for livestock because "a single plant product may constitute a significant portion of the animal diet. For instance, 50-75 per cent of the diet of most domestic animals consists of field corn. Therefore, a change in nutrient or toxicant composition that is considered insignificant for human consumption may be a very significant change in the animal diet".

There is already evidence that 'green revolution' crops - introduced in the 1960s and 1970s as a solution to world hunger - have caused themselves problems due to their changed nutrient composition. High-yielding green revolution crops have been found not to take up essential minerals from the soil, so that when the proportion of calories provided to people by these crops rose, their intake of key micro-nutrients fell. If GM crops are to 'feed the world', they must be subject to the most stringent safety testing. Monsanto and Aventis have used substantial equivalence to argue that significant differences in composition are of little importance in the diet as a whole. But this is not an appropriate way to assess the safety of GM foods for the world's poor and vulnerable.

SAFETY OF GM FOODS

One of the most commonly voiced concerns about the safety of GM foods is the possibility that changes in nutritional value, or even toxic effects, will be produced in the GM food as an unexpected consequence of genetic modification. Establishing the substantial equivalence of a GM crop or food is reliant upon being able to identify any changes that might have occurred. At the moment, tests only look at what is already known to be present in the food. They measure the levels of individual, known components, such as protein, starch or fat, to see if there is any change. This cannot detect anything new or unexpected.

As a recent OECD report on the safety testing of GM foods pointed out, "analysis based on single compounds as a screening method for unintended effects of genetic modification, has its limitations with regard to (unknown) anti-nutrients and natural toxins". As was also noted in the OECD report, information about the interactions of the many different compounds found in food is limited and inevitably leads to problems in analysing the results. Often there is only limited information on what constitutes a normal level of naturally occurring toxins. This means that there is a limit to any understanding of what the impacts that a change in the levels of any one of the many compounds found in foods might have in terms of food safety.

In the most common approach, levels are deemed acceptable if they fall within the range of values published in the literature for all the varieties of that particular crop grown around the world. But there is no scientific basis for this as a definition of safety. Indeed, it is not even clear whether such information is reliable. When discussing this kind of data at an OECD meeting, it was noted that "the quality and level of this information was variable and did not always inspire confidence".

For the vast majority of foods, even those which are eaten widely throughout the world and regarded as staples, understanding of the impact of the food and its constituents on health is fairly limited. This leads to difficulties when assessing any changes to a food caused by genetic modification. For example, in a report produced by the ACNFP on insect-resistant maize it was stated that "there are no inherent toxic or antinutritional factors present in maize". Yet recently it has been suggested that linoleic acid, which is a major fatty acid of maize, may be linked to immune system suppression when it forms a major part of the diet, as is the case for many poor people in developing countries. The extent of this effect is still unclear, but changes in linoleic acid content which have previously been considered to be insignificant could have a serious impact in those areas where maize is a staple food. Changes in fatty acid profiles, including linoleic acid content, have been found in GM maize varieties produced by AgrEvo, Monsanto and Pioneer Hi Bred.

Allergic reactions

Genetic modification frequently makes use of proteins from organisms which have never previously been an integral part of the human food chain. Allergic reactions are caused by an excessive immune reaction to particular substances (known as allergens) in susceptible people. Exposure to the allergen, such as by eating a food, inhaling pollen or by touch, triggers the production of an excess of immunoglobulin E (IgE) class antibodies. The development of an allergy may take time, but once it has developed there may be no safe level of exposure for the affected person. Allergies only affect a small proportion of the population, up to 2. 5 per cent (although up to 25 per cent of the population self-report allergies), but the consequences can be severe, even deadly.

In addition, the incidence of severe reactions appears to be rising, with the number of cases of anaphylactic shock increasing five-fold in the period 1985 to 1995. The number of food substances implicated is on the increase, and there is also a reported increase in allergy to pollens. There is concern that the introduction of novel proteins into GM foods could be a source of new allergies for sensitive members of the population and further increase this rising trend. The approach taken by Monsanto scientists for the assessment of the allergenicity of novel proteins was published in 1996,and similar approaches have been taken by other companies.

If the novel protein originates from a plant or other organism known to cause allergies, then the allergenicity of the novel food can be assessed by using blood sera from people with an allergy to the source of the novel protein. This can be followed by tests on human volunteers if required. In a famous example, soya modified with a gene from the brazil nut, in order to increase the content of the amino acid methionine, was found to elicit an allergic reaction in people who were sensitive to brazil nuts. Because the source of the novel protein was known to cause allergies, it was relatively straightforward to test the GM soy, and its development was subsequently halted.

But, as noted by the Monsanto scientists, if the novel gene and protein comes from an organism that has never before been eaten as food, then it is of unknown allergenicity and the assessment is far more difficult. The scientists proposed that in this case, the novel protein should be compared to known allergenic compounds to see if there are any similarities. In fact, most of the novel proteins introduced into GM crops so far have come from sources of unknown allergenicity, such as bacteria or non-food plants, and this comparison approach has been widely used.

One of the assumptions often used in companies' assessments of the allergenicity of a novel protein is that allergens are usually found in large amounts in allergy-causing foods. But this is not always true - in the case of apples the main allergen makes up only 0. 3 per cent to 1. 5 per cent of total protein. In fact, in the case of cow's milk allergy, it has been found that some allergic reactions appear to be caused by proteins that are only present in trace amounts, such as immunoglobulins or bovine serum albumin. So, although many of the novel proteins found in novel foods are only present in small quantities, this is no guarantee of their safety.

In the case of novel proteins from non-food sources, their amino acid sequence is established and this is compared to databanks of the amino acid sequences of known allergens. Around 200 allergens from food, pollens, mites and animals have been sequenced, and it is estimated that if eight amino acids are found in sequence in common with any of these allergens then there could be cause for concern. While this is an effective means of eliminating some potential allergens, it is limited by the fact that many allergens are not yet categorised. For example, beta-lactoglobulin, an allergen from cows milk, is not in the databanks and would not be identified using amino acid comparison either.

The fact that the amino acid sequence of a novel protein does not show any sequence similarities with the allergens listed in these databanks does not guarantee that it is not an allergen. In addition, it is not just the amino acid sequence but the shape of the protein (essentially how the chain of amino acids is folded) which determines its allergenicity. Many allergens of animal origin share a similar structure, but this cannot yet be used as a tool for identifying new allergens as the shape of only a very few allergens has been determined.

Novel proteins are frequently examined for resistance to digestion. This is based on the assumption that the allergic reaction is caused by the whole protein and that allergens resist digestion. Experiments have shown that many food allergens resist digestion for up to one hour, while normal food proteins are broken down in less than one minute. In most assessments, the novel protein is tested in vitro using extracted or synthetic gastric juices from a range of animals and humans. The rate at which the protein breaks down is seen as an indicator of its allergic potential.

Inappropriate testing has been used as an assurance about the safety of GM foods for the last ten years. In May 2000, the OECD Task Force for the safety of Novel Foods and Feeds produced a report on the safety assessment of GM foods. The Task Force was made up of officials from national governments around the world - 25 countries and the European Commission sent participants. The report of the task force says: " Classical animal feeding trials are inadequate, due to the difficulty of feeding animals adequate doses of the test food in their diet. Increased research efforts and new techniques are needed to develop alternative safety assessment techniques for whole foods, especially in the areas of immunotoxicology, gut toxicology, molecular biology and plant physiology. "

If there is to be any kind of confidence in GM foods, better alternatives must be found. Increasingly, researchers are looking at in vitro tests using cells as a means of screening for effects. Tests exist which use liver cells or cells from the lining of the gut to assess the toxicity of chemicals, and some researchers have already suggested that such tests could be used for GM foods. The use of cell-based tests for examining possible toxic effects is already common in the pharmaceutical industry for the initial screening of new drugs and the same approach could be used for GM foods.

It is also important to examine what happens once the components in the GM food are absorbed into the body. So for example, various GM foods are in development which contain the snowdrop lectin. Researchers who examined the effect of this chemical on human white blood cellsfound the chemical could bind to the blood cells, indicating a potential toxic affect. Research of this kind, which looks beyond simple animal feeding experiments, must be an essential part of the safety assessment procedure for GM foods. But isn't. Without doubt, GM foods present a fresh challenge for food-safety testing and new techniques must be developed if their safety is to be assessed reliably. There is no easy way out or short cut for producers of GM foods, as far as consumers are concerned.

Chapter 6

Imperatives and Challenges for Rice Biotechnology

Rice, Oryza sativa L., is the dominent crop in Asia. It is the staple food for more than three billion people or over half the world's population. It provides 27 per cent of the dietary energy supply and 20 per cent of dietary protein intake in the developing world. Grown in at least 114 mostly developing countries, rice is the dominant crop in Asia where it covers half of the arable land used for agriculture in many countries. Moreover, it is the primary source of income and employment for more than 100 million households in Asia and Africa.

The Asian continent, where 56 per cent of humanity including 70 per cent of the world's 1.3 billion poor people lives, produces and consumes around 92 per cent of the world's rice. Over 50 per cent of the 840 million people suffering from chronic hunger live in areas dependent on rice production. About 80 per cent of the world's rice is produced in small farms, primarily to meet family needs, and poor rural farmers account for 80 per cent of all rice producers. Nine of the top ten rice producing countries in 2003, viz., China, India, Indonesia, Bangladesh, Vietnam, Thailand, Myanmar, the Philippines, and Japan are in Asia.

China and India combined account for more than half of the world's rice area and, along with Indonesia, consume more than three-fourths of the global rice production. However, less than seven per cent of the world's rice production is traded internationally and, with this small marketable surplus, prices fluctuate widely with droughts, floods, and typhoons. But more to being the world's most popular staple, rice has been cultivated by mankind for more than 10,000 years. Hence, it provides a symbol of global unity and cultural identity for many countries where rice cultivation is practically intertwined with religious observances, festivals, customs, folklore, and traditions.

CHALLENGE IN RICE PRODUCTIVITY

The Green Revolution beginning in the 1960s ushered in an era of high rice productivity, with yields doubling or tripling from 1.9 tonnes per hectare (t/ha) in many Asian countries (Figure 1). Between 1966 and 2000, the rice production growth of 130 per cent, from 257 million rice-growing countries tonnes in 1966 to 600 million tonnes in 2000, outpaced the population growth of 90 per cent in low income countries. The average per capita food availability was 18 per cent higher in 2000 than in 1966. About 84 per cent of rice production growth has been attributed to the use of modern technologies such as rice varieties that are semi-dwarf, early maturing, non-photoperiod sensitive and can, therefore, be planted more than once a year, and responsive to nitrogen fertilizer.

More than 2,000 modern varieties, with resistance and tolerance to biotic and abiotic stresses, have been commercially released in 12 countries of South and Southeast Asia over the last 40 years. As a consequence, rice production cost per unit output was reduced by 20-30 per cent. This translated to reduced rice prices at the consumer level from about US$450 per metric tonne (mt) unmilled rice in the early 1950s to less than US$300 per mt by 1999, with world market rice prices decreasing by 80 per cent over the last 20 years. In Asia, however, demand for rice is projected to increase by 70 per cent over the next 30 years, driven primarily by population growth that, excluding

China, is expected to increase by 51 per cent. It is estimated that the Asian population will increase from 3.7 billion in 2000 to 4.6 billion in 2025.

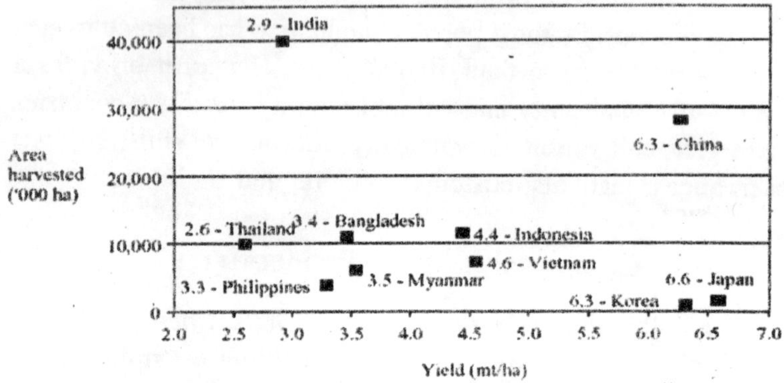

Figure 1: Area harvested and yield levels (mt/ha) in major Asian

In the Philippines, for example, the population is expected to reach 107 million by 2025 and 65 per cent more rice, relative to present levels, would need to be produced to keep up with demand by a population continually growing by 2.3 per cent each year. This translates to a required sustained year on year rice production growth of 3 per cent. Given that annual rice production growth rates have been decelerating to less than 2 per cent per year, and the land frontier, the primary source of growth in recent years is closing, major technological progress has to be achieved in the next two decades to avoid importation. Against a backdrop of decreasing land, labour, and water and rising prices of production inputs, the challenge to further increase rice productivity is indeed enormous.

Rice production and post-production processes in Asia are severely compromised by pests, diseases, and physiological and environmental factors. The rice crop, for example, is the world's single largest market for agrochemicals, consuming around US$3.7 billion annually, with agrochemical costs and crop losses amounting US$ tens of billions per year. Tungro, the most destructive viral disease in Southeast Asia, for example, results in crop losses worth more than US $1.8 billion annually. Moreover, rice cultivation per se is restrained by resource constraints, among the most important of which are the projected scarcity of water and scarcity of land.

Technological progress is required to increase crop water productivity of rice and rice crop productivity under fragile environments such as the rainfed lowlands and uplands. Biophysical constraints account for substantial yield losses in Asia (Table 1). In the irrigated ecosystem, yield loss due to technical constraints accounted for 20 per cent (962 kg/ha) of the average yield, with soil-related problems being the most significant. On the other hand, yield loss due to technical constraints accounted for 33 per cent of average yield in the rainfed lowland and flood-prone ecosystems, with submergence being the most important, while it was more than 40 per cent of the average yield in the upland ecosystem, with drought being the most significant.

Overall, all technical constraints caused a total yield loss of about 23 per cent or 833 kg/ha in Asia, with abiotic constraints being more important than biotic constraints for all ecosystems. Climate-related constraints like submergence, drought, and cold resulted in yield losses that ranged from 227 kg/ha (20 per cent of average yield) for the upland to 429 kg/ha (28 per cent of average yield) for the flood-prone ecosystems. Yield losses due to pests and diseases, on the other hand, were most significant in the rainfed ecosystem while, that due to weeds were most important for the upland environment.

Role for Rice Biotechnology

Agricultural biotechnology in Asia has been recognized as having the potential to:

(i) increase crop and animal productivity;

(ii) improve nutritional quality;

(iii) broaden tolerance of crops for drought, salinity, and other abiotic stresses; and

(iv) increase resistance of crops to pests and diseases.

In the case of rice biotechnology, the Rockefeller Foundation (RF), in the process of developing its International Programme on Rice Biotechnology or IPRB, identified the top-20 priority traits for biotechnology research intervention, balancing research costs vis a vis the benefits from expected increases in rice productivity or value.

Guided by this prioritization, the RF in the mid-1980s supported a 17-year programme that laid the scientific foundation for 'rice biotechnology' as we know it to-day. At about the same time, Asian national agricultural research and extension systems (NARES) began building biotechnology capacity. Among the important accomplishments of the IPRB were:

— the generation of the first DNA molecular marker map of rice;

— the regeneration and transformation of rice;

— the use of rice pest genomic information to understand host-plant resistance;

— discoveries that changed the way rice geneticists viewed breeding objectives such as insect resistance, abiotic stress tolerance, and hybrid rice;

— the discovery of rice's pivotal genomic position in the evolution of cereal species;

— the transfer of the resulting biotechnologies to institutions in rice-producing and - consuming countries; and

— the strengthening of both physical and human resources in cooperation with national and international rice research systems in Asia, Africa, and Latin America.

The latter involved international collaborative research-cum-training that successfully linked emerging national rice biotechnology efforts directly to advanced research institutes in

Table 1: Yield loss due to technical constraints in the rice ecosystems of Asia.

Constraints	Irrigated (kg/ha)	Rainfed lowland (kg/ha)	Flood-prone (kg/ha)	(kg/ha)	Average for Asia Upland Loss	Loss (kg/ha) (%)
Biotic						
Diseases	69	146	18	70	83	3.1
Insects	108	166	16	65	110	2.3
Other pests	29	88	21	120	52	1.4
Abiotic						
Water	400	288	429	227	358	9.9
Soil	356	75	13	80	229	6.4
Total	962	763	496	563	833	23.1
Loss as % of yield	19.6	33.2	33.1	40.2	23.1	

the United States, Europe, Japan, and Australia, resulting in the training of more than 400 rice scientists, primarily from Asia, in advanced laboratories around the world. Beginning 1993, the Asian Rice Biotechnology Network (ARBN) was also formed, with IRRI as coordinator, and this facilitated research collaborations amongst several Asian rice breeding programmes with a primary objective of developing disease resistant varieties through the application of DNA marker technology.

APPLICATIONS OF RICE BIOTECHNOLOGY

With increased activities on rice biotechnology beginning in the mid-1980s, rice gradually became the 'model monocot plant' in molecular genomics research, eventually becoming the first food crop for which complete genome sequence became available. Progress achieved in the application of biotechnology for rice improvement has been in two major areas-the use of molecular markers for identifying and incorporating favourable genes within the rice species, and the use of transgenic technologies to incorporate traits for herbicide tolerance, biotic stress resistance, abiotic stress resistance, and nutritional value into rice.

Use of molecular markers. The development of the first rice molecular map in the late 1980s sped up molecular genetics research in rice. Among the early molecular marker applications for rice improvement were the:

— construction of dense genetic maps using different populations;

— tagging and/or introgression of major genes and those underlying quantitative traits, referred to as quantitative trait loci (QTL);

— high-resolution characterization and fingerprinting of germplasm;

— assessment of the diversity of germplasm pools; and

— map-based gene cloning.

Molecular markers offered great potential for increasing the precision and speed of rice breeding as, among other advantages over phenotypic markers, they provided the ability to

— screen breeding populations regardless of growth stage;
— screen for traits that were extremely difficult, expensive, or time consuming to score phenotypically; and
— distinguish the heterozygous condition without need for progeny testing.

Molecular markers provided geneticists with powerful tools to dissect the inheritance of economically important traits in rice, many of which are quantitatively inherited and complex in nature. Thus, studies dealing with QTL were carried out such as those involving tolerance to a variety of environmental stresses including drought, seedling vigour, submergence, salinity, and mineral deficiencies or toxicities. These traits were considered primary targets for molecular marker-aided selection (MAS) as breeding for them using conventional techniques often proved to be difficult.

MAS or the selection of traits based on the presence or absence of a molecular marker or markers in lieu of phenotype has already received a lot of emphasis in rice. The development of simple and less costly marker systems based on the polymerase chain reaction (PCR) such as the simple sequence repeats or SSRs contributed greatly to the use of MAS in various laboratories in developing countries. For example, at PhilRice, the NARES for rice in the Philippines, MAS studies are conducted to develop varieties resistant to bacterial blight, including the pyramiding of two to three bacterial blight resistance genes in a common genetic background, both for inbred and hybrid rice breeding.

Gene pyramiding is expected to provide durable resistance against rice insect pests and diseases and attempts in this direction have been tried early on for bacterial blight and rice blast diseases and the insect brown planthopper. Introgressing genes from wild relatives into cultivated rice has also been accomplished with the aid of molecular markers, such as the bacterial blight resistance gene from O. longistaminata, and the yield traits from O. rufipogon. Markers have also been used to minimize the linkage drag that occurs in wide crosses and to obtain the desired recombinants in fewer generations during backcrossing. Whole-genome, marker-based selection allows for new opportunities to unravel and makes efficient use of genetic variation both in cultivated rice and its wild relatives.

One of the most significant developments aided by the use of molecular markers in rice was the map-based cloning of Xa 21 and its subsequent use in developing varieties with broad spectrum resistance to bacterial blight. Starting with the genetic mapping using RFLP makers of the Xa 21 locus in 1990, the gene was cloned using map-based cloning techniques and a bacterial artificial chromosome library by 1995. By 1997, the gene had been pyramided with other Xa genes using PCR-based MAS and, by 1998, Xa 21 had been transformed into elite lines with field trials conducted in China, India, and the Philippines by 1999.

By 2000, a hybrid rice parental restorer line had been improved through MAS, resulting in resistant hybrid rices under field conditions. The IRRI-coordinated Asian Rice Biotechnology Network that was supported by ADB and the RF played a key role in developing capacity for marker-aided analyses of pathogens and host plant resistance in several national breeding programmes. This network approach was found essential for the sharing of resources and providing sustained training in the adoption of new biotechnology tools and genetic knowledge in individual breeding programmes of different NARES in Asia. As a result of ARBN activities, elite or commercial rice lines with multiple disease resistance genes have been developed in several participating countries (Table 2).

Transgenic Technologies

No genetically modified (GM) or transgenic rice has yet been commercialized in Asian countries. However, two GM rice, both involving herbicide tolerance, have already passed regulatory approval processes in the US: the Liberty-Link™ rice of Aventis Crop Science (now Bayer CropScience) involving phosphinothricin (PPT) herbicide tolerance, specifically glufosinate ammonium, and the CLEARFIELD™ rice involving imidazolinone herbicide tolerance from BASF Inc.. Ten trials in 11 hectares and 12 trials in 45 hectares were conducted in 2002 and early 2004, respectively, 90 per cent of which involved Monsanto. To indirectly gauge the extent to which the use of GM technology had so far advanced in rice, utilized information on patent applications and classified these into the areas of:

— herbicide tolerance;

— biotic stress resistance;

— abiotic stress resistance; and

— nutritional traits.

Until 2002, 307 patents on rice biotechnology from 404 different groups had been filed. The largest number of patents was held by DuPont/Pioneer (68), followed by Monsanto (33), Syngenta (32), Bayer (19), public sector institutions in Japan, and Japan Tobacco. Amongst various traits, herbicide tolerance has been the major focus for the private sector. In the US, Monsanto and Bayer were responsible for 80 per cent of GM rice field trials, primarily addressing herbicide tolerance. Other countries where herbicide tolerant GM rice has been field tested include Italy, Brazil, Argentina, and Japan, and possibly China.

Biotic stress resistance, on the other hand, has been the primary focus for public sector research institutions including those in Asia. Specific traits being worked on using GM technologies include resistance to bacterial blight using the gene Xa21, rice blast, rice hoja blanca virus, rice tungro spherical

Table 2: Marker aided selection (MAS)-improved varieties and their corresponding increase in yield developed by research teams from Asian NARES

Country	Background commercial/ Yield standard	Released (R) / Near-release (NR)	Yield (t/ha)	Gain over yield standard (%)
Philippines	IR64	AR32-19-3-2 (NR)	5.1	0
	IR64	AR32-19-3-3 (NR)	6.7	31.4
	IR64	AR32-19-3-4 (NR)	6.1	19.6
	BPI Ri10	AR32-4-3-1 (NR)	6.0	17.6
	BPI Ri10	AR32-4-58-2 (NR)	6.5	27.5
	PSB Rc28	Yield standard	5.1	-
Indonesia	IR64	Angke (Bio-1) (R)	5.4	20.0
	IR64	Conde (Bio-2) (R)	5.4	20.0
	IR64	Yield standard	4.5	-
India	PR106	IET17948 (PR106-P2) (NR)	8.2	22.4
	PR106	IET17949 (PR106-P9) (NR)	7.9	17.9
	PR106	Yield standard	6.7	-
China	Zhong 9A/Zhonghui 218	Hybrid Guofeng No. 2 (NR, R)	7.8	11.4
	II-3A/Zhonghui 218	Hybrid II You 218 (NR, R)	8.3	18.6
	Shanyou 46	Yield standard	7.0	-

virus, rice yellow mottle virus, rice ragged stunt virus, the brown planthopper, and yellow stem borer, the latter, using Bt technologies, being the closest to commercialization.

For abiotic stress tolerance, transgenic rice plants that produce trehalose at 3-10 times the normal rate, resulting in tolerance to drought and/or salinity have been developed by introducing the ots A and ots B genes for trehalose biosynthesis from Escherichia coli into rice. In China, Chinese researchers have developed several GM rice varieties that are resistant to the country's major rice pests and diseases, such as the stem borer, bacteria blight, rice blast fungus and rice dwarf virus. Significant progress has also been made with drought - and salt-tolerant varieties of GM rice, which have been in field trials since 1998.

Perhaps one of the most promising, despite controversies, application of transgenic technology in rice has been the development of Vitamin A-enriched rice, popularly known as Golden Rice™ due to the slight yellow color conferred on the rice endosperm in transgenic grains. Vitamin A is considered as absolutely essential for children and women of child bearing age and, worldwide, nearly 134 million children are at risk for diseases related to Vitamin A deficiency (VAD), including some 3.1 million preschool age children who suffer from eye damage, and nearly 2 million children under 5 years of age that die each year from diseases linked to persistent VAD.

In Southeast Asia alone, 5 million children become at least partially blind every year due to VAD. Golden Rice™ has the potential to improve the supply of Vitamin A in the human diet, thereby alleviating the suffering and death of millions of people, especially those who cannot afford diet diversification (ISAAA 2004a). New Vitamin Aenriched materials with up to ten times more the level of pro-Vitamin A are now in the pipeline, including several popular Asian indica varieties such as IR64. Reported to involve "clean" events, without cross-border transfers or antibiotic markers, the new materials are being readied for backcrossing and stability and field testing in 2004, while vitamin A absorption and bioavailability tests are underway or planned in the Philippines, China, and the USA.

Another promising use of transgenic technology to improve human nutrition is in combating iron deficiency, one of the most widespread micronutrient deficiencies worldwide. Deficiency to iron in the human diet affects about 3.5 billion people worldwide and could result in illnesses such as anemia, heart problems, and neurological disorders. The ferritin gene from Phaseolus vulgaris has already been introduced into rice, resulting in the doubling to tripling of the iron content in the rice endosperm, even after polishing the grain. To improve the bioavailability of iron, since it is usually in complex with phytic acid, the genes from Aspergillus fumigatus encoding a thermotolerant phytase protein and the endogenous cysteine-rich metallothionein-like protein were also introduced into rice resulting in a 7-fold increase in cysteine level and a 130-fold increase in phytase level in the resulting transgenic plants.

Many NARES in rice-growing countries of Asia are actively involved in the use of transgenic technologies, encouraged by government policies supporting biotechnology research, and endowed with universities and agricultural research institutes with biotechnology research capacity. In a recent study done by the International Food Policy Research Institute, 209 transformation events were reported to have already been done in 76 scientific institutes in 16 countries. Of these, 109 (52 per cent) were done in 7 Asian countries, viz., China (30), Indonesia (24), India (21), the Philippines (17), Thailand (7), Pakistan (5), and Malaysia (5).

Of these countries, however, only the Philippines had so far approved the commercial release of a transgenic food crop - a Bt-enhanced corn. Although the highest number of transformation events for any crop was reported for rice (17.7 per cent), followed by potatoes (11.0 per cent), maize (8.6 per cent), and papayas (6.2 per cent), a GM rice variety has yet to be commercialized in Asia.

In the Philippines, both IRRI and PhilRice have on-going rice biotechnology programmes employing molecular marker and transgenic technologies, as well as other more conventional techniques such as in vitro culture and wide hybridization. With the Philippine government declaring a policy supportive of biotechnology research, the use of biotechnology is embedded as a strategy for achieving the goals set by the irrigated lowland, direct seeded, rice for fragile environments, and hybrid rice multidisciplinary R&D programmes of PhilRice. GM technology, in particular, is being used to improve high-yielding varieties, including NPTs and hybrid rice parental lines.

The Philippine focus is on the tungro, sheath blight, blast, and bacterial blight diseases, as well as on the insect stemborer, and tolerance to salinity. Genes procured from laboratories around the world and modified for Agrobacterium tumefaciens-mediated transformation, are being used to generate transgenic plants. A number of transgenic plants that contain chitinase and glucanase genes have already been produced and tested under controlled screenhouse conditions. Moreover, PhilRice has conducted the first and only contained field trials for any GM rice in the Philippines where the transgenic IR72 plants containing the Xa21 gene for bacterial blight resistance showed complete resistance against nine Philippine races of the pathogen.

On the other hand, transgenic plants with the pin2 gene are being developed to improve stemborer resistance, while a coat protein gene from the rice tungro bacilliform virus is being used in A. tumefaciens-mediated transformation. PhilRice is also a member of the Golden Rice™ Network and undertook backcrossing activities on the now discarded original Golden Rice™ materials. To hasten the availability of vitamin-A enriched rice to consumers, PhilRice hopes to continue its active participation in this network that involves other Asian countries such as Indonesia, Vietnam, India, Bangladesh, and China, as well as partners in developed countries such as the US, Germany, UK, and Switzerland. Already, guidelines are on the national testing of GM rice prior to commercialization are being prepared.

To hasten public acceptance of biotechnology, in general, and GM rice, in particular, PhilRice, along with other Philippine government agencies, has also been spearheading a massive public education campaign using the tri-media as well as various public fora involving the government, private, NGO, and religious sectors. Advances being made in the field of functional genomics will provide ample scope to further increase yield, build plant protection, improve nutrition, and enable rice to grow using less water and in adverse environment. The availability of the complete rice genomic sequence offers a lot of opportunities to further understand the natural genetic variation and the effects of alleles, and their interactions, for traits that are important in rice breeding, using specific genetic backgrounds and under specific environments.

At IRRI, for example, biotechnologists are systematically assessing the array of phenotypes resulting from the disruption of putative gene sequences in mutants, near-isogenic lines, permanent mapping populations, and elite and conserved germplasm through a functional genomics initiative. However, these gene discovery and allele mining efforts would require the annotation of the rice genome and the subsequent build up of databases and information resources. The use of information and communication technology and bioinformatics, such as the IRIS and GeneFlow databases, should make the mounting information more easily accessible to scientists, especially breeders in rice-growing countries.

Other work in progress include studies designed to transfer the C4 photosynthetic pathway and leaf anatomy genes of maize to C3 rice in order to improve the rice plant's radiation use efficiency while reducing transpirational water loss and N fertilizer requirement, and studies aiming to more deeply understand the genetic variation for drought tolerance using genomics and bioinformatics tools in order to identify the exact genes involved. Another promising area is the genetically engineering of N2 fixation capacity into rice and, in this direction, attempts to engineer the nif-regulon into the chloroplast genome have been made with the idea of not having to make rice totally independent of external nitrogen supply, but to provide additional nitrogen during the grain filling period to rescue the photosynthetic apparatus for a longer period of time.

Over the medium- and long-term, the apomixis research that earlier had been started at IRRI, would need to be vigorously pursued using biotechnology in order to capture and made available to resource poor farmers the benefits of heterosis. With over 27 million tonnes of rice worth US$5560 million lost annually due to pests such as the stem borer, sheath blight, and bacterial blight, GM rice solutions are nearing commercialization to help offset this yield loss. Ready for field evaluation are Bt/ hybrid rice in China and India; XA rice in China, Philippines and India; Golden rice in the US, the Philippines and India; and herbicide tolerant rice in Spain, USA, and China. For commercialization by 2005-06 will be Bt rice in China, and Golden Rice and Ferritin or high- iron rice by 2007-08.

There is a need for many Asian rice-growing countries to develop clear and time-bound national agenda for biotechnology R&D and commercialization. For example, in the Philippines, a technology explicit and market driven national development agenda, which recognizes the role of science and technology in promoting economic development and facilitating trade is desired, along with corresponding increased investments in R&D. This could be achieved by developing a collaborative scheme that shall bridge the academe and science and technology community with industry, and having both well tuned to market demands. There is also a need to emphasize the role of the government in formulating technology policies and plans, as well as funding of research and development projects; the role of academe in identifying what problems need solving, and the role of the private sector in investing research that could meet the country's immediate needs.

In the case of India, for the first time, a specific impetus has been given to the advancement of biotechnology. In the country's 2005-05 budget, the Minister for Finance emphasized that science and technology, including biotechnology, "will receive priority and will be provided with additional funds." A specific provision was given for companies doing scientific R&D approved by the Department of Scientific and Industrial Research before April 1, 2004 to be entitled to 100 per cent deduction of profits for 10 years. The Federation of Indian Chambers of Commerce and Industry noted that this will favour small and medium biotechnology companies who can channel savings to augment R&D activities.

The Agricultural and Development Economics Division (ESA) of the Economic and Social Department of the United Nations' Food and Agricultural Organization (FAO) showed that most of the research institutions in China, Cambodia, and Indonesia are in the initial states of developing biotechnology research capacity. The proportion of biotechnology expenditures to the total agricultural research expenditures is very small, and mainly from the public sector.

For instance, countries like China, India, Indonesia, Malaysia, Philippines, Argentina, Brazil and Bulgaria have made significant investments in biotechnology research and regulation. Experience with genetically modified organism (GMO) testing for a wide array of locally important traits, and commercialization in these countries is growing, and government support programmes and policies actively encourage biotechnology R&D.

In China, scientists have released a report urging the central government to allow the commercial planting of genetically modified (GM) rice. They note that GM rice technologies are technically mature and are ready to be commercialized. GM products include several rice varieties resistant to China's major rice pests, including those that can resist the stemborer by using Bacillus thuringiensis (Bt), delta endotoxin and cowpea trypsin inhibitor CpTI genes, a protease inhibitor rice, a planthopper and bacterial leaf blight resistant rice using the Xa21 gene, and fungus-resistant rice. The country currently has the largest field for GM rice trials. An estimate of about 25-30 per cent of China's plant biotechnology investments are spent on GM rice programmes.

China has increased its budget for research and field trials of GM rice since 2001. Its biotech budget for 2001-2005 is $1.2 billion, a 400 per cent increase compared with 1996-2000. About $120 million out of the current budget is devoted to GM rice programmes. The Chinese Government has become the world's second-largest spender on plant biotechnologies, next only to the United States and the country is expected to launch at least 10 GM rice field trials by 2005 and release Bacillus thuringensis (Bt), cowpea trypsin inhibitor gene, and Xa21 gene GM rices that year. Already, field trials in Hunan and Fujian provinces showed that GM rice boosted yields by 4 to 8 per cent, and allowed an 80 per cent reduction in pesticide use.

Setting the biotechnology R&D priorities may be argued that most of the earlier rice biotechnology activities, particularly in the public sector and on transgenic technology applications, were more sciencedriven and researcher-driven than attuned to the needs of ordinary farmers. It is important to note that for GM rices to be useful, at least in the immediate term, and to gain fast acceptance amongst resource poor farmers, it is best that they be derived from varieties already widely grown and suited to specific agroenvironments.

Ranged against the challenges confronting rice cultivation in most Asian rice growing countries today, it is clear that international as well as NARES biotechnology research should focus, on one hand, on the most important rice diseases and pests, and physiological and environmental factors that reduce rice productivity and quality, and on the other hand, on increasing rice yield potential. An example of trait prioritization for rice biotechnology research in Asia, aimed at delivering the greatest impact on the lives of poor rice farmers and consumers, while being complementary to conventional rice improvement efforts, has been put forward by Hossain et al..

To ensure relevancy of biotechnology R&D agenda, a bottomup approach needs to be implemented in the crafting of R&D priorities, where farmers' and other stakeholders' needs and concerns are adequately addressed. Such approach should benefit from the rich indigenous knowledge of local farming communities on specific rice production constraints while facilitating public acceptance and ensuring the trickling down of benefits from biotechnology-derived products. The extent to which biotechnology may improve food security does not settle the question of its relevance to quality of life; it is essential to move stakeholder involvement upstream in setting research priorities and to fairly share benefits through appropriate infrastructure and ownership arrangements.

There remains an imbalance on R&D investments on rice biotechnology that tend to favour developed countries, thus impacting on the potential of biotechnology to boost agriculture in the developing world and alleviate the plight of resource-poor rice farmers in Asia. The concentration of biotechnology R&D in developed countries and the limited private sector effort in developing countries, particularly in Asia, has raised concerns over the economic concentration of biotechnology in favour of developed countries and multinational companies. As the private sector would be unlikely to undertake rice biotechnology research based primarily on the pressing needs of resource poor farmers, due to difficulty in recovering costly investments, there is need for significant publicsector funding initiatives if benefits are to reach resource-poor farmers in developing countries in order for them to develop pro-poor biotechnology R&D agenda.

Furthermore, public research products would have to gain similar approval as those developed by the private sector if transgenic research products and their associated potential benefits are to reach the poor. While Asian scientists have demonstrated the capacity to successfully undertake biotechnology R&D relevant to the needs of resource-poor farmers, the desired phenotypes have been few when compared to traits being developed by multinational firms and advanced research institutes in the developed world.

One bright spot, however, has been the case of Thailand that established the National Center for Genetic Engineering and Biotechnology (BIOTEC) in 1983. BIOTEC has supported biotechnology R&D in six areas, including the improvement of disease resistance in rice, particularly rice blast. This disease affected 200,000 hectares of rice in Thailand in 1993, causing serious economic loss and resulting in government intervention to assist disease-struck farmers worth about US$10 million.

Since then, BIOTEC has supported research for the molecular genetic characterization of local blast isolates and mapping of resistance blast genes, with focus on aromatic varieties for Thailand's export rice market. In 1999, BIOTEC also provided US$3.7 million to fund the 'Rice Genome Project Thailand', particularly for the sequencing of rice chromosome 9 that contains a QTL for tolerance to a very important concern of Thai rice farmers-submergence. Most rice growing countries in Asia, with the exception of China and India, however, have yet to launch similarly focused government initiatives on rice biotechnology R&D.

In China, investments on public sector biotechnology research has risen dramatically to $1.2 billion for 2001-2005, a 400 per cent increase over 1996-2000 levels, with about $120 million allocated for transgenic rice R&D. With field testing on various transgenic rices already going-on since 1998, with 53 hectares planted in 2003, China is poised to becoming the first country in the world to commercialize transgenic rice. Importance of collaborations. Given the varying biotechnology research capacities of rice-growing NARES and the limited resources allocated for biotechnology research in the public sector, unintentionally encouraged by the phasing out of the IPRB of the Rockefeller Foundation, the constraints in NARES R&D budgetary allocations, and the reduction of funding support for international agricultural research centers (IARCs) such as IRRI, the need for biotechnology R&D practitioners to collaborate has become paramount.

Collaborations need to be pursued at the individual, institutional, governmental, bilateral, regional, and international levels to ensure not only that the highest returns for R&D investments are attained, but also to facilitate regulatory approvals and biotechnology product commercialization. At the national level, the creation of a national coordinating body such as BIOTEC in Thailand should provide a mechanism for increasing efficiency in the use of limited national R&D budgetary allocations through the avoidance of research duplications and sharing of in-country research capacities.

In the Philippines, the recent creation of a Crops Biotechnology Center under the Department of Agriculture augurs well for the development of a unified biotechnology R&D agenda for the country's priority crops, including rice, and the optimization of the use of limited manpower and logistical resources. On the other hand, a regional collaboration approach, as exemplified by the ARBN should be able to develop a biotechnology R&D agenda focused on the shared needs of rice farmers in the region and, where possible, human pool, scientific, and financial resources or, alternatively, parcel out the research portfolio as was done in the rice genome sequencing initiative.

One type of formal collaboration that has yet to be explored involves bilateral arrangements between countries. In the development of transgenic technologies, these South-to-South collaborations would be facilitative to the build-up and sharing of common approaches, genes, germplasm, regulatory trials, and biosafetyrelated information. The already established broad-based regional cooperation, such as the Association of Southeast Asian Nations (ASEAN) and the Asia Pacific Economic Conference (APEC) should be tapped to support these regional biotechnology undertakings. At the international level, programmes that help rice scientists from developing countries to train, further hone their capacities, and maintain ties with advanced laboratories at IARCs and developed countries need to be supported.

For instance, India's Department of Biotechnology (DPT) and the United States for International Development have signed a letter of intent to initiate enhanced cooperation in agricultural biotechnology research and development. The partnership will pursue agri-biotech research projects including technology development, technology diffusion, biosafety, and related policy activities. In addition, joint workshops, conferences, scientific exchanges, and training of scientists will also be done. The programme will increase the "range of safe and environmentally sound technological options to producers and consumers of agricultural products".

Role of IARCs and Private Sectors

Without formal, dynamic, and synergistic interfaces between the public and private sectors, much of the benefits of crop biotechnology will not reach those who need them the most. The sharing of information and experiences across sectors is crucial to facilitate the flow and process that technologies undergo from the laboratory to the farm. With many NARES still not fully enabled to undertake, solely by themselves, activities spanning the whole biotechnology research, development, and commercialization spectrum, IRRI and similar international institutions will still need to play their roles as technology and knowledge providers, as well as builders and enhancers of biotechnology capacities of rice-growing NARES.

Of particular importance for IRRI is the provision of strategic research outputs that already several NARES in Asian countries are capable of transforming into applications and products. These could include protocols, gene constructs, and markers for traits relevant to local problems but prohibitively expensive for NARES to develop by themselves. Alternatively, IRRI should be able to complement its strategic research programme with a product development thrust, focusing on biotechnology-derived advanced breeding lines and varieties, with traits that have common high relevance amongst Asian countries. The product development portfolio could include varieties tolerant to drought, of high nutritional value, and resistant to major diseases such as tungro and bacterial blight.

The role of IRRI as facilitator in the transfer of useful technologies and products amongst NARES through the sharing of technologies, knowledge, and experiences need to be strengthened. Equally important is its role in facilitating the formation of effective NARES/public sector and private sector collaborations, for NARES to be able to access private sector-held intellectual property (IP) on rice biotechnology processes and products. IRRI can also serve as a clearinghouse for IP-protected technologies from both the public and private sectors that NARES can easily access.

Training support by IRRI and similar institutions for NARES should now also include those designed to advance NARES capacity on the science and management of biotechnology, IPR, biosafety and food safety regulation, and international negotiations. As Cantrell and Hettel (2004) argued, with IRRI's strengths, it can serve as the unbiased broker and facilitator amongst the rice NARES, advanced research institutions, and the private sector. Other international organizations, such as the FAO, can help expedite rice biotechnology progress in Asia by promoting and supporting networking mechanisms such as the South-to-South cooperation model.

They can also help in developing and supporting infrastructure for public-good agricultural research, providing knowledge and training to NARES researchers, enabling interactions amongst biotechnology stakeholders through dialogues and similar fora, facilitating access to relevant IP, sensitizing policymakers on biotechnology-related issues, and assisting governments in the crafting of biotechnology-related policies. As the primary source of GM crops continues to be the private sector, technology transfer between private and public sectors, in terms of products as well as experiences on regulation, commercial development, and release of GM crops, would greatly benefit NARES. This technology transfer can be facilitated by private foundations such as the International Service for the Acquisition of Agri-Biotech Applications (ISAAA).

Impacts of Intellectual Property Rights (IPR)

The impact of IPR on biotechnology research is often imbedded in discussions on public and private sector partnerships. There is a need to balance the fact that on one hand public sector institutions, due to limited resources, cannot fully avoid accessing private sector-held IP during the development of its own products and, on the other hand, the private sector has to avail of IPR protection to be able to protect its investments and commercial interests as well to be able to share their IP with other sectors without fear of exploitation. The development of the Golden Rice™ is a case in point.

In total, 70 IPRs and technical property rights (TPRs) belonging to 32 different companies and universities were used in product development and for which 'freedom-to-operate' situations had to be applied for in order for NARES to begin using Golden Rice™ in further breeding and in de novo transformation activities using locally adapted varieties. Several modalities, however, are still open to the public sector to be able to access genes and technologies from the private sector. These include direct purchase of genes and technologies, licensing, the fact that patents have time limits, confidential agreements, and the purchase of genes for incorporation into local germplasm.

New types of IP agreements have also evolved, such as the donation of IP facilities and 'humanitarian' use type agreements as were done with Golden Rice™ with the threshold for humanitarian versus commercial use being a $10,000 income from the technology. The agribusiness giant Syngenta recently announced that it would donate new Golden Rice seeds and lines to the Golden Rice Humanitarian Board, including scientific results of the first field trials, as well as the technology, rights, and research results. As the issue of IPR is likely to become increasingly important, the capacities of governments or the science sectors of many developing countries to understand, deploy, and negotiate regarding biotechnology need to be strengthened.

Rice biotechnology practitioners in Asia need to be trained on the details of modern IPR systems and on negotiating with institutions and companies for the purpose of accessing IP, and applying for IP protection. Alternatively, research institutions could establish IP units, not only for negotiating with other institutions and sectors, but also for the IP registration of their own biotechnology processes and products.

National biosafety committees in developing countries have made impressive progress in the drafting and implementation of biosafety regulations for the importation and testing of transgenic crops, with regulations for field tests already in place in rice-growing countries such as China, India, Thailand, and the Philippines. A looming issue, however, revolve around the compliance costs for regulatory approval that could prohibit many developing country institutions. In the various studies cited by Atanassov et al., annual compliance costs, including initial greenhouse and field screening, field testing for environmental impact, and food safety, but excluding technology development costs, ranged from US$140,000 for a virus-resistant papaya in Brazil to US$830,000 for a virus resistant potato in South Africa.

For rice, an annual regulatory compliance cost of US$680,000 was estimated for a virus resistant variety in Costa Rica covering tests on molecular characterization and epidemiology, transgenic field trials, biosafety, IPR, food safety deployment, and gene flow. Given reduced NARES budgets, this could pose a major hurdle in the commercialization of rice biotechnology products from the public sector. It is hoped, however, that as knowledge and experience is gained by regulatory agencies, approval costs may decrease, both by reducing the number of required tests, and by shortening the length of experimentation.

The latter would also avoid the risk of biotechnology products becoming irrelevant to farmers' needs due to approval delays. In this regard, continuous training of personnel from regulatory bodies of developing countries on new biotechnology developments and approaches is necessary for them to make educated recommendations, as is envisioned in the Cartagena Biosafety Protocol. A well functioning regulatory system can hasten public acceptance of biotechnology products by instilling public confidence that the risk assessments conducted are carefully done, science-based, and, therefore, safe. While international harmonization of standards may be required, there is also need for contract-sensitivity that is appropriate to the place in which a technology will be applied.

Policy initiatives are necessary to accelerate investments by technology holders and adoption by the farming communities in Asia. The Indian government, for example, is formulating new policies to boost investments and research in the local biotechnology sector. These include promoting the speedy approval of GM crops, and funding and infrastructure support for public-private partnership programmes plant biotechnology, among other areas. The new policies will also provide the framework for research and business institutions, and illustrate the trade and investment guidelines for the newly emerging biotech sector. A group of experts will also be set up to suggest models for public-private partnerships in the biotech sector.

The biotechnology department will invest the creation of innovation centers with the existing academic and research institutions. To date, Delhi University has been identified to be the first center to receive the funding. Other cities such as Hyderabad, Pune, Chennai, Ahmedabad, and Lucknow will also be the focus of the development. Similarly, in Taiwan, Premier Yu Shyi-kun pledged his commitment to establish a Biotech Industry Strategy Consulting Committee, which would consolidate and integrate the country's biotech-based promotion organizations and research institutes.

Aside from policy initiatives, transparent protocols must also be established and international guidelines followed. There is also the need to establish a world-class intellectual property rights regime for biotechnology inventions and the protection of plant varieties. In addition, concurrent and transparent trials for biosafety for both new and released events are necessary and a determined drive against illegal genetically modified seeds trade is important.

Biosafety and Food Safety

Biotechnology is a very powerful tool that can be used in the developing world to grow more rice in an environmentally friendly manner. It can improve food production by making farming more efficient. The benefits that biotechnology confers upon the environment include the reduction in the use of agrochemicals and the preservation of presently uncultivated and marginal lands and their concomitant biodiversity due to increases in productivity in the more favorable environments. To sustain the rice agriculture resource base and avoid environmental disturbance, it is important to match new genes and biotechnology-derived varieties to the conditions of the target environments.

As commercialized GM crops by 2003 in the developing world was largely limited to insectprotected cotton in Argentina, China, India, Mexico, and South Africa, there is still limited experience on food safety assessments to draw from for a food crop such as rice. Among developing countries, only three have approved a single transgenic event in a food crop (soybean in Brazil, Czech Republic and Uruguay; and maize in the Philippines), two have approved two events (soybean and tomato in Mexico; soybean and maize in South Africa) and one (Korea) has approved three events (one in soybean and two in maize). Therefore, the sharing of experiences and knowledge from the food safety assessments done in these countries should be valuable for developing countries with rice biotechnology products in the pre-commercialization stages.

As rice is a food crop, and per capita rice consumption could vary both in-country and among countries, from less than 100 kg/yr in countries like China and India to over a 200 kg/year in countries like Myanmar, careful food safety experimentation must be done in the case of nutrient-enhanced GM rice to remove potential health dangers related to over dosages, if any, or alternatively, effective GM rice deployment strategies need to be developed.

Communication

Communication plays a vital role in disseminating information. It is a powerful tool in influencing man's decision. Media plays a crucial role in framing public understanding of science. Appreciation of traditional knowledge is essential for science to be communicated successfully. The message will be understood and better appreciated if there is an understanding of local knowledge. It is important to understand cultural values, and respect traditional knowledge to successfully communicate science. Scientists, journalists and others involved in the communication of science should take into account cultural factors and local knowledge in their work.

The adoption and use of biotechnology and genomics should be approached within the context of conventional practices. Rather than view agricultural biotechnology as the wave of the future, we should regard it as part of a mélange of the new and old, balanced appropriately to meet local needs. The point was emphasized that agricultural biotechnology would be more readily adopted, and food security would be more attainable as would environmental sustainability, if it were blended back into conventional practices in order that valueadded benefits would accrue alongside maintenance of traditions.

There are good examples of public education campaigns on biotechnology being done in Asian NARES. In Thailand, the Biotechnology Alliance Association (BAA) has been set up to educate the public about biotechnology applications, including genetically modified crops. Science based information will be provided to the public and several fora will be organized to discuss the benefits and risks of GM crops so that stakeholders can make intelligent decisions relative to biotechnology and its use. The potential for biotechnology to result in economic, environmental, and social benefits in India is enormous and promotion of the country's first hand experience in this area, through knowledge sharing activities, could serve as a powerful example for other developing countries.

Chapter 7

Evaluation of Food Safety

Food manufacturers are required by law to ensure the safety and quality of their products regardless of the source or identity of the ingredients. Traditional foods are viewed by the Food and Drug Administration as "safe" based on a long history of use. The consuming public also views traditional foods as safe. However, many traditional foods contain naturally occurring toxins that can present hazards to consumers under some circumstances of exposure.

Fortunately, in most circumstances, these naturally occurring toxins are present in concentrations that are not hazardous to consumers ingesting typical quantities of the food prepared under typical conditions. Also, some traditional foods are allergenic to some consumers, even though they are safe for the vast majority of consumers. New foods produced through conventional breeding or introduced into the marketplace from other parts of the world are not required to undergo any type of safety assessment.

They are assumed to be safe because they are comparable to other varieties or because they have been safely consumed in other parts of the world. In fact, these newly introduced foods may contain numerous unique components that are not individually or collectively assessed for safety. In contrast, products derived through rDNA biotechnology are assessed for safety before their introduction into the food marketplace. Food manufacturers also must ensure the safety and quality of products that contain ingredients derived from rDNA biotechnology. FDA provided a general outline for the safety assessment of rDNA biotechnology-derived food products based on risk analysis related to the characteristics of the products.

All of the existing foods produced using rDNA biotechnology have undergone a rigorous science-based safety assessment focusing on the characteristics of the products, especially the unique components. While this practice has been voluntary in the United States, FDA announced in May 2000 that it intends to propose a premarket notification system for rDNA biotechnology-derived foods that would make this unofficial policy into a regulatory requirement.

SUBSTANTIAL EQUIVALENCE

In the safety assessment of rDNA biotechnology-derived foods, it is helpful to compare the new plant variety to its traditional counterpart because the counterpart has a history of safe use as a food. The concept of substantial equivalence effectively focuses the scientific assessment on potential differences that might present safety or nutritional concerns. Substantial equivalence is not an absolute determinant of safety per se, since compositional changes in an rDNA biotechnology-derived food may have no impact on the safety of the food.

However, substantial equivalence provides a process to establish that the composition of the plant has not been changed in such a way as to introduce any new hazards into the food, increase the concentration of inherent toxic constituents, or decrease the customary content of nutrients. For example, high-oleic-acid soybean oil from rDNA biotechnology-derived soybeans has an oleic acid concentration that falls outside the range typically found in soy oils. From a scientific perspective, this food is nevertheless considered safe, based on scientific knowledge about the safety of oleic acid, a common fatty acid in foods.

A determination of substantial equivalence considers the intentional and unintentional effects of genetic modification, and includes an evaluation of phenotypic and compositional characteristics. With respect to food safety, substantial equivalence involves the quantitative assessment of the concentration of inherent constituents in the modified food, compared to the often wide range typically found in its traditional counterpart, under similar food production conditions.

Most food sources (e.g., soybeans, corn) are exceedingly complex mixtures that vary widely in composition, so it is necessary to consider all of the factors that determine the normal range of variation Key constituents measured include nutrients, such as proteins, fats, carbohydrates, vitamins, and minerals, as well as inherent antinutritional factors, toxins, and allergens. The breadth of technology used to measure these constituents is evolving rapidly, with new methods available to assess the integrity of metabolic pathways and to measure secondary metabolites, functional proteins, and gene expression at the molecular level.

The Food and Agriculture Organization of the United Nations (FAO) and the World Health Organization (WHO) considered the concept of substantial equivalence: A comparative approach focusing on the determination of similarities and differences between the genetically modified food and its conventional counterpart aids in the identification of potential safety and nutritional issues and is considered the most appropriate strategy for the safety and nutritional assessment of genetically modified foods.

The Consultation was of the view that there were presently no alternative strategies that would provide a better assurance of safety for genetically modified foods than the appropriate use of the concept of substantial equivalence. Nevertheless, it was agreed that some aspects of the steps in safety assessment process could be refined to keep abreast of developments in genetic modification technology. The concept of substantial equivalence was developed as a practical approach to the safety assessment of genetically modified foods.

It should be seen as a key step in the safety assessment process although it is not a safety assessment in itself; it does not characterize hazard, rather it is used to structure the safety assessment of a genetically modified food relative to a conventional counterpart. The Consultation was satisfied with the approach used to assess the safety of the genetically modified foods that have been approved for commercial use. The Organization for Economic Cooperation and Development (OECD) examined the safety of novel foods and feeds. It concluded that:

> Safety assessment based on substantial equivalence is the most practical approach to address the safety of food and food components derived through modern biotechnology.

In its 1992 policy on foods derived from new plant varieties, FDA employs the concept of substantial equivalence by focusing on the characteristics of the food product. Foremost, this policy on food products from new plant varieties is intended to be applied regardless of the derivation of the plant, i.e., through conventional breeding or rDNA biotechnology methods.

These include a substance that is completely new to the food supply, an allergen expressed in an unusual or unexpected circumstance, changes in the concentrations of major dietary nutrients, and increased concentrations of antinutritional factors and toxins inherent to the food. Although the FDA policy does not specifically use the term substantial equivalence, the absence of the characteristics mentioned above would lead to the conclusion that a food from a new plant variety is substantially equivalent to its traditional counterpart.

Safety of Gene Product

Under FDA's current policy, as a starting point, the

characteristics of the product are assessed, including the nucleotide sequence of the DNA of the genetic material that is used for plant transformation. This procedure provides important information on the encoded protein(s), regulatory elements controlling expression, and the presence or absence of additional potential coding sequences within the DNA. Although all extraneous non-coding DNA may not be identified, it can be minimized to very small segments.

This level of detail cannot ordinarily be determined for new plant varieties produced in conventional ways such as hybridization. Thus, the FDA policy contemplates that the structure and function of proteins encoded by the gene(s) introduced into plants will be understood in considerable detail. This information is used to assess the level of any potential risk, both of the introduced protein and of other products that may be produced or altered by the presence of the introduced protein. An additional factor is the source of the gene.

Any potential safety concerns associated with the source organism would serve to focus the safety assessment of the rDNA biotechnology-derived plant and the products derived from that plant. For example, if a gene were obtained from a source that produced a known allergen, the proteins encoded by the introduced DNA would have to be assessed to demonstrate that this DNA did not encode an allergen. Safety of Introduced Genetic Material. The initial step in a safety assessment is full characterization of the genetic construct being inserted.

This step includes identifying the source of the genetic material to establish whether it originates from a pathogenic, toxin-producing, or allergenic source. Parameters measured include the size of the genetic construct that is inserted into the plant genome, the number of constructs inserted, the location of insertion, and the identification of genetic sequences within the construct that allow for its detection and expression in the plant. The genetic material transferred is composed of DNA. All food, rDNA biotechnology-derived or otherwise, contains DNA.

Individuals consume large quantities of DNA when eating conventional foods. The DNA introduced using rDNA biotechnology represents only a tiny fraction of the total DNA consumed when the food is eaten, and transfer of genes from rDNA biotechnology-derived plants to mammalian cells is extremely unlikely. Since DNA occurs in all foods, it is not subject to a safety evaluation. It is well-established that DNA is rapidly digested in the gastrointestinal tract, and there is no evidence of DNA transfer from foods to human intestinal cells or gut microorganisms.

Any plant DNA that might be found in human tissues is likely to be a small, non-functional fragment resulting from centuries of consumption and does not imply that plant foods are unsafe. Moreover, the likelihood of transfer of rDNA segments from foods produced using rDNA biotechnology is far less than for DNA from conventional foods simply because the novel DNA is less than 1/250,000 of the overall amount consumed. Earlier rDNA biotechnology-derived foods were based on the use of selectable marker genes that confer resistance to an antibiotic.

The presence of marker genes per se in food would not constitute a safety concern. FAO/WHO recently reconsidered the issue of antibiotic resistance marker genes and again found there is no evidence that the markers currently in use pose a health risk to humans or domestic animals. Still, genes that confer resistance to drugs with specific medical use or limited alternative therapies should not be used in widely disseminated rDNA biotechnology-derived foods. Following extensive examination, FDA decided to permit the use of kanamycinresistance genes in the development of rDNA biotechnology-derived tomatoes, oilseed rape, and cotton for food and feed use and permitted these crops in food and feed.

The DNA for kanamycin resistance was not different from other rDNA in its digestibility and does not pose a food safety concern. The marker gene used to confer kanamycin resistance was the neomycin phosphotransferase, type II gene (NPTII). The NPTII protein is rapidly degraded, like other dietary proteins, when subjected to conditions which simulate mammalian digestion. This protein has also been tested in acute toxicology studies at levels more than one million times the level that would be consumed by people eating food from rDNA biotechnology-derived plants.

Finally, the transformation of intestinal bacteria by kanamycin resistance from plants is negligible, with a calculated theoretical maximum of less than 1 in 100,000 compared to bacterial transfers of resistance. Thus, this protein poses no food safety concerns. There is no inherent danger presented by the presence of the antibiotic resistance markers used in earlier rDNA biotechnology-derived foods. These marker genes, such as the NPTII gene, do not present a food or feed safety concern and are not considered to be either toxic or allergenic.

The risk that the use of antibiotic resistance genes could lead to a transfer of antibiotic resistance and reduced efficacy of antibiotics is extremely small, because it would require a series of events, each of which is highly unlikely. Moreover, if such a move did occur, antibiotic selection would be needed to make the newly resistant strain a common one. FDA's 1992 policy also contemplates that, once the genetic construct has been fully characterized, an assessment of the safety of the gene product will be conducted.

The gene product is the protein, often an enzyme, that is produced by the newly introduced gene(s) and is present in the rDNA biotechnology-derived food or food ingredient, e.g., the protein expressed in Bt corn, encoded by genes from Bacillus thuringiensis (Bt), that confers pesticidal specificity for lepidopteran insects.

Safety evaluations typically include identification of the composition and structure of the gene product; a quantification of the amount of gene product expressed in the edible portion of the food; a search for similarity to known toxins and antinutritional factors, allergens, and other functional proteins; a determination of the thermal and digestive stability of the gene product; and the results of both in-vivo and in-vitro toxicological assays to demonstrate lack of apparent allergenicity or toxicity.

Unintended Effects of Genetic Modification

From a safety perspective, unintended effects of genetic modification have been speculated to manifest as the unintended expression of some unknown or unexpected toxic or antinutrient factor, or the otherwise unintended enhanced production of known toxic constituents. However, based on the knowledge gained to date from the multitude of foods derived from rDNA biotechnology, there is no scientific evidence of the occurrence of such unintended effects.

Given the more precise and predictable nature of genetic change accomplished through rDNA techniques as compared to the random genetic changes observed in conventional breeding, such unintended effects would be considered less likely in foods derived from rDNA biotechnology. Furthermore, these effects have been observed infrequently in the many thousands of crosses involving conventional crop breeding. In such cases, the source of the toxic constituent can typically be traced back to a related species used in conventional cross-breeding manipulations.

These toxins are present in all potatoes, and new potato cultivars are routinely screened for glycoalkaloid content. The unusually high glycoalkaloid content in Lenape was attributed to the use of the wild, nontuber- bearing Solanum chacoense in its parentage. Interestingly, Lenape is a parent of Atlantic, a current potato variety with a glycoalkaloid content typical of the range for edible potatoes.

Allergenicity of Foods

Food allergies involve abnormal immunological responses to substances in foods, usually naturally occurring proteins found in commonly allergenic foods such as peanuts, milk, and seafood. Allergic reactions can be manifested by symptoms ranging from mild cutaneous or gastrointestinal symptoms to life-threatening anaphylactic shock reactions. Virtually all food allergens are proteins, although only a small fraction of the proteins found in nature are allergenic.

Since genetic modifications involve the introduction of new genes into the recipient plant and since these genes would produce new proteins in the improved variety, the potential allergenicity of the newly introduced protein should be a key component of the safety assessment process. An assessment of the potential allergenicity of rDNA biotechnology-derived foods typically follows the decision-tree process outlined by the International Food Biotechnology Council (IFBC) and the Allergy and Immunology Institute of the International Life Sciences Institute (ILSI).

This strategy focuses on specific scientific criteria, including the source of the gene(s), the sequence homology of the newly introduced protein(s) to known allergens, the immunochemical reactivity of the newly introduced protein(s) with immunoglobulin E (IgE) antibodies from the blood serum of individuals with known allergies to the source from which the genetic material was obtained, and the physicochemical properties, e.g., digestive stability, of the introduced protein.

Several other criteria, including the level of expression of the newly introduced protein(s) in the edible portions of the improved variety and the evaluation of the functional category for the introduced protein, were suggested for addition to the IFBCILSI allergenicity assessment strategy. The first step of the allergenicity assessment involves the classification of the source of the genetic material as either commonly allergenic, less commonly allergenic, or of unknown allergenic potential.

Eight foods or food groups, including milk, eggs, fish, crustacean shellfish, peanuts, soybeans, tree nuts, and wheat, are well accepted as commonly allergenic; these eight foods or food groups account for more than 90% of all food allergies in the world. More than 160 other foods have been described to cause allergic reactions, and would be classified as less commonly allergenic. However, many of the genes that have been and will be used to produce rDNA biotechnology-derived foods are obtained from sources with no history of allergenicity as foods.

Certainly, if the source contains well known environmental allergens, e.g., ragweed that contains common ragweed pollen allergens, then the allergenicity of newly introduced protein(s) from such sources must be carefully evaluated. The approaches to allergenicity assessment vary according to the nature of the source of the transferred genetic material.

If the genetic material is obtained from a known allergenic source, either commonly or less commonly allergenic, and the encoded protein is expressed in the edible portion of the rDNA biotechnology-derived food, then the protein must be considered to be an allergen unless proven otherwise. In such situations, the next step in the allergenicity assessment is a determination of the immunoreactivity of the newly introduced protein with IgE antibodies from the sera of individuals allergic to the donor organism.

The blood serum can be tested for reactivity with the purified protein or extracts of the genetically modified food using immunoassays. If a sufficient number of test sera are used as advocated in the decision tree approach, the allergenicity of the introduced protein can be determined with a high degree of confidence. However, if negative results are obtained in the immunoassays, the rDNA biotechnologyderived food or extracts of that food should be tested further using in-vivo skin-prick tests, double-blind, placebocontrolled food challenges, or digestive stability assessments as advocated by the IFBC-ILSI decision tree.

If the immunoassays and these other tests, as appropriate, are negative, then the likelihood that the rDNA biotechnology-derived food contains an allergen would be quite small. The most difficult assessment occurs when genes are obtained from sources with no history of allergenicity, such as viruses, bacteria, or non-food plants. The likelihood that the proteins derived from such sources of DNA will be allergens is not very high, since most proteins in nature are not allergens.

Additionally, many of these proteins will be expressed in the rDNA biotechnology-derived food at very low levels, while allergic sensitization is more likely to occur to the major proteins that exist in foods. The key features of the allergenicity assessment for such foods involve a comparison of the amino acid sequence of the introduced protein with the amino acid sequences of known allergens and the digestive stability of the introduced protein.

While the combination of these two criteria provides reasonable assurance that the introduced protein has limited allergenic potential, the ideal approaches to the application of these two criteria have been debated, and the desirability of adding other criteria for the allergenicity assessment of such products has been advocated. The criterion of amino acid sequence homology to known allergens is a logical and increasingly powerful approach. The amino acid sequences of more than 300 known allergens are available for comparative purposes.

The IFBC-ILSI strategy defines significant sequence similarity as a match of at least eight contiguous, identical amino acids based on the minimal peptide length needed for T-cell binding, which is a necessary prelude to allergic sensitization; this approach is clearly limited in that it cannot identify discontinuous or conformational epitopes that are dependent on the tertiary structure of the protein. Others have suggested that the definition of significant sequence homology be modified to a minimal peptide length of less than eight contiguous, identical amino acids.

While this criterion (amino acid sequence homology to known allergens) is clearly useful, international agreement must be sought on its application. Known food allergens tend to be quite stable to digestive proteases with the exception of the pollenrelated food proteins that cause oral allergy syndrome. Thus, digestive stability can be used as a criterion for the assessment of the allergenic potential of the introduced proteins. Both simulated gastric and intestinal models of mammalian digestion are advocated for such assessments.

While the usefulness of this criterion is apparent, consensus is needed on the ideal protocols for assessment of digestive stability. It is recognized that novel proteins may exist that are stable to digestion but will not become allergens. Additional testing is needed to assess the allergenic potential of such proteins. The development of additional criteria and additional tests to use in the assessment of the allergenicity of rDNA biotechnology-derived foods would be advantageous in cases where the gene is obtained from sources with no history of allergenicity.

As mentioned, the level of expression of the introduced protein and the functional category of the introduced protein could be used as additional criteria. In addition, the development of suitable animal models for the prediction of the allergenic potential of the introduced proteins is anticipated in the future. While several animal models appear to be promising, none has been sufficiently validated for its routine use in the assessment of the allergenicity of rDNA biotechnology-derived foods.

The existing decision-tree approach has already been applied in the assessment of the allergenicity of rDNA biotechnology-derived foods. The enzyme introduced into glyphosate-tolerant soybeans has no sequence homology to known allergens and is rapidly digested in simulated mammalian digestion systems. Similarly, several of the Bt proteins used in insect-resistant crops and the proteins produced by common marker genes are rapidly digested in simulated mammalian digestion systems.

A high-methionine protein introduced into soybeans by the transfer of a gene from Brazil nuts to correct the inherent methionine deficiency in soybeans was shown to bind to IgE from the sera of Brazil nut-allergic individuals and to elicit positive skin-prick tests in some of these patients. This protein was thus identified as the major allergen from Brazil nuts that had not previously been characterized.

As a result, commercial development of this particular soybean variety was discontinued. Clearly, the assessment of the allergenicity of rDNA biotechnology-derived foods should be a key component of the overall safety assessment process in all cases. A useful strategy has been developed for such assessments, although this strategy should be viewed as dynamic and new approaches and criteria should be added once they are validated and accepted.

Conventional Counterparts

Recombinant DNA-derived biotechnology foods without conventional counterparts need to be evaluated on a case-bycase basis and would be subject to some types of toxicity assessments, depending on the nature of the modification. This situation has not yet arisen with rDNA biotechnology derived foods, although at some point it undoubtedly will. When it does, the situation will raise variety of issues that will need to be addressed in a scientifically based, flexible manner.

Whole foods are complex mixtures of chemical components characterized by wide variations in composition and nutritional qualities, and are not well suited for traditional toxicological studies designed to assess individual chemical entities. The testing of whole foods-rDNA biotechnology-derived or conventional-in animal feeding studies, for example, is limited by factors such as the animal's qualitative and quantitative feeding preferences and the levels of nutritional and antinutritional factors and other substances that are present.

When one researcher attempted to ascertain the toxic threshold for an rDNA biotechnology-derived tomato by feeding rats freeze-dried tomato extract, the experiments were limited to the human equivalent of 13 tomatoes a day by negative effects of inorganic compounds, such as potassium, that are present in rDNA biotechnology-derived and conventional tomatoes alike. But, as noted by MacKenzie, "Toxicologists still said we hadn't fed them enough to get a meaningful result."

Another limitation is that animal toxicity tests are seldom sufficiently sensitive to distinguish differences between the toxicity of a new variety and its conventional counterparts. Indeed, most foods will produce adverse effects in long-term animal feeding studies when fed in high proportions of the diet, regardless of the nature of production. There is no scientific justification for requiring long-term feeding studies for rDNA biotechnology-derived foods, and that such studies would be unlikely to provide meaningful information in the great majority of cases.

FAO/WHO concurred, finding that the practical difficulties in the application of conventional toxicology studies to whole foods preclude their use as a routine safety assessment technique. Thus, given a hypothetical rDNA biotechnology-derived food without a conventionally derived counterpart, animal studies would need to be designed to address specific nutritional or toxicological concerns. However, these studies would need to be carefully designed to avoid or minimize the limitations discussed above that are associated with the testing of whole foods or major food constituents.

For example, toxicological studies could be used to examine the potential for acute, chronic, carcinogenic, genotoxic, reproductive, and teratogenic effects of components or fractions of concern in a food derived from a new plant variety. A complete assessment would also include pharmacokinetic data regarding absorption, distribution, metabolism, and excretion of the new product or a novel component thereof. By focusing toxicological examination on carefully selected fractions or components of a food derived from a new plant variety, and excluding major components of no concern, it may be possible to reduce or eliminate the difficulties associated with testing whole foods.

The assessment of macronutrient substitutes or other major food constituents should follow a tiered approach, whereby the physical and chemical properties of the constituent are determined, in addition to its potential to disrupt or alter nutrient uptake. Initial predictive effect studies would dictate the physiologically relevant endpoint determinants of subsequent in-vitro and in-vivo studies.

Further, the choice of animal model for any such invivo studies would have to be carefully considered for relevance when applying results to humans. Without precedence, the above discussion outlines a proposal which seems best calculated to provide the data needed for a persuasive showing of safety. Clearly, such novel foods without conventional counterparts, when they do become available, will need careful testing, evaluation, and regulatory scrutiny using a flexible process that contains case-by-case adaptation based on the novel nature of the issues presented.

SCIENTIFIC CONSENSUS ABOUT SAFETY

The Human Food Safety Panel reviewed available information about the safety of rDNA biotechnology-derived foods and found that there is striking congruence in the conclusions and recommendations of various international scientific groups that have considered the issue. Most significant conclusions and recommendations include (1) there is no evidence of the existence of unique hazards, either in the use of rDNA biotechnology techniques or in the movement of genes between unrelated organisms, and (2) the risks associated with the introduction of rDNA biotechnology-derived organisms are the same in kind as those associated with the introduction of unmodified organisms and organisms modified by other methods.

No conceptual distinction exists between genetic modification of plants and microorganisms by classical methods or by molecular techniques that modify DNA and transfer genes. The same physical and biological laws govern the response of organisms modified by modern molecular and cellular methods and those produced by classical methods. Recombinant DNA methodology makes it possible to introduce pieces of DNA, consisting of either single or multiple genes, that can be defined in function and even in nucleotide sequence.

The risks associated with biotechnology are not unique, and tend to be associated with particular products and their applications, not with the production process or the technology per se. In fact biotechnology processes tend to reduce risks because they are more precise and predictable. The health and environmental risks of not pursuing biotechnology-based solutions to the nation's problems are likely to be greater than the risks of going forward.

Biotechnology has a long history of use in food production and processing. It represents a continuum embracing both traditional breeding techniques and the latest techniques based on molecular biology. The newer biotechnological techniques, in particular, open up very great possibilities of rapidly improving the quantity and quality of food available. The use of these techniques does not result in food which is inherently less safe than that produced by conventional ones.

Food safety considerations regarding organisms produced by techniques that change the heritable traits of an organism, such as rDNA technology, are basically of the same nature as those that might arise from other ways of altering the genome of an organism, such as conventional breeding. While there may be limitations to the application of the substantial equivalence approach to safety assessment, this approach provides equal or increased assurance of the safety of food products derived from genetically modified organisms as compared to foods or food components derived by conventional methods.

There is no strict dichotomy between, or new categories of, the health and environmental risks that might be posed by transgenic and conventional pest-protected plants and that the properties of a genetically modified organism should be the focus of risk assessments, not the process by which it was produced. The committee concluded that "[w]ith careful planning and appropriate regulatory oversight, commercial cultivation of transgenic pestprotected plants is not generally expected to pose higher risks and may pose less risk than other commonly used chemical and biological pest-management techniques."

The safety of rDNA biotechnology- derived foods has been extensively reviewed by a number of scientific organizations, at the national and international level. The use of rDNA biotechnology in itself has no impact on the safety of such foods. Foods derived using rDNA biotechnology are subject to rigorous and systematic scientific evaluations under existing principles of food safety-far more than are routinely applied to the products of traditional breeding. Thus, the level of field testing and premarket review for food safety provide assurance that foods derived from plants and microorganisms through rDNA biotechnology are at least as safe as existing foods, and are consistent with all existing standards of food safety.

LABELING OF rDNA BIOTECHNOLOGY-DERIVED FOODS

The United States food labeling requirements directly relevant to the labeling of recombinant DNA biotechnologyderived foods, including constitutional limitations on the government's authority to regulate food labeling and specific case studies relevant to labeling rDNA biotechnology-derived foods.

Food and Drug Administration Requirements and Policies: Generally speaking, the Food and Drug Administration (FDA) has authority over food labeling, and the Federal Trade Commission (FTC) has authority over food advertising. Except for meat and poultry products regulated by the U.S. Department of Agriculture (USDA), the federal law governing the labeling of food generally is the Federal Food, Drug, and Cosmetic Act (FFDCA).

The FFDCA is administered by FDA. Under this statute, FDA regulates food labeling through a series of requirements that are intended to assure that information of significance about a food product is provided and that food labeling is truthful and not misleading. "Labeling" is defined in the FFDCA as "written, printed, or graphic matter (1) upon any article or any of its containers or wrappers, or (2) accompanying such article". Thus, "labeling" includes-but is not limited to-the "label" that is physically attached to the immediate container of foods in package form.

Physical attachment or proximity of the material to the product is not required for the material to be considered "labeling" for purposes of the statute. A product that was sold separately from the product was nevertheless "labeling" for purposes of the statute because the product and the booklet "were parts of an integrated distribution scheme". Material that is not regulated as labeling by FDA will be regulated as advertising by FTC. At the most basic level, the FFDCA and its implementing regulations specify that certain information is required on the labels of almost all foods.

These label requirements are intended to assure provision of information that is fundamental to the description of the food or the operation of the food safety regulatory system. Examples of these label requirements are the common or usual name (or other name) of the food; net contents statement; an ingredient listing for food products made from more than one ingredient; name and place of business of the manufacturer, packer, or distributor; and nutrition labeling.

Constitutional Constraints: All statutory labeling requirements, their implementing regulations, and FDA labeling policies must satisfy constitutional requirements. The principal constitutional consideration in food labeling matters is First Amendment constraint of government labeling regulation. Until the 1970s, advertising or labeling restrictions were viewed as purely economic regulations that did not implicate the First Amendment. Indeed, until the late 1970s, the Supreme Court had excluded commercial speech from the coverage of the First Amendment.

The government may restrict commercial speech if (1) the speech is either misleading or concerns an unlawful activity, or if (2) the asserted governmental interest in support of the restriction is substantial, (3) the restriction directly advances the government's substantial interest, and (4) the regulation is not more extensive than is necessary to serve that interest.

The First Amendment protects both the right to speak and the right not to speak. The constitutionally protected right not to speak, the compelled speech doctrine, is clearly established in Supreme Court precedent. Indeed, the Supreme Court has suggested that compelling someone to speak involuntarily is an even more serious constitutional matter than preventing speech. The regulation of food labeling involves both the commercial speech and the compelled speech doctrines. The courts have not articulated a "compelled commercial speech" doctrine.

False or Misleading Statements: Beyond these fundamental label requirements and constitutional constraints discussed above, the food processor is generally at liberty to make use of label or labeling space in the manner it deems fit, provided that the label or labeling is not false or misleading. The FFDCA deems a food to be misbranded if "its labeling is false or misleading in any particular". The prohibition of misleading labeling is the objective of many of the specific labeling requirements of the FFDCA, as well as the basis for most FDA regulation of voluntary labeling statements.

If a statement, picture, or other representation on the label or labeling of any food product is false or misleading, the food is misbranded regardless of the importance of the representation to the consumer. The Supreme Court has held that it is not necessary to show that anyone was actually misled or deceived, or that there was any intent to deceive, in order to find that a product is misbranded under the FFDCA. Other courts have stated that the test is not the effect of the label on a "reasonable consumer" but on "the ignorant, the unthinking, and the credulous" consumer.

The prohibition on false or misleading labeling statements reaches far beyond patently false claims. Statements that, while not false, are misleading are also prohibited. For example, a "cholesterol-free" claim for broccoli suggests that particular broccoli is cholesterol-free, while ordinary broccoli is not cholesterol-free. Thus, the claim is misleading, since ordinary broccoli does not contain cholesterol. To reinforce this interpretation, the FFDCA explicitly prohibits a claim that states the absence of a nutrient unless the nutrient is usually present in the food. To avoid being misleading, FDA permits the claim "broccoli, a cholesterol-free food," but not "cholesterol- free broccoli" .

Just as labeling statements may be misleading because of what they say or imply, they may be misleading by virtue of what they do not say. In determining whether a food labeling statement is misleading, FDA and the courts take into account the extent to which the labeling fails to reveal any material facts. There is neither a statutory nor a reg- ulatory definition of "material fact," and the term has not been elaborately defined by the courts. Instead, determinations of whether or not a fact is material are made on a case-by-case basis, with an extensive body of precedents.

Generally, if a new or modified food is significantly different from its conventional counterpart in composition, nutritional value, or safety, the difference in the food would be considered a material fact. For example, if a new processing technique resulted in a significant decrease in the nutrient content or change in flavor, color, or other valued characteristic of a food, a label statement would be required to inform consumers of that material fact. Absent a label statement disclosing a material fact about a food, the presentation of the food would be misleading.

So the FFDCA prohibition on false or misleading labeling may effectively require that a label include a disclosure of the material fact. While some FDA disclosure requirements are imposed to provide for safe use of food ingredients or to provide consumer warnings, many disclosure requirements are imposed to clarify or explain an otherwise misleading label statement. For example, FDA decided that a statement of the percent reduction is necessary to clarify a claim like "reduced fat". The agency determined that consumers would likely be confused unless the magnitude of the reduction was specified.

Federal Trade Commission Requirements: FTC regulates food advertising under the Federal Trade Commission Act (FTCA), which is similar in structure to the FFDCA. The FTCA generally prohibits "deceptive acts or practices in commerce". It prohibits false advertising that is likely to induce the purchase of foods, and declares such false advertising to be prohibited as "deceptive acts or practices". The term "false advertising" is defined as advertising that is "misleading in a material respect". These FTCA provisions are similar to the FFDCA provisions on labeling that is false or misleading.

To determine if advertising is false or deceptive, FTC examines whether and to what degree the information in the advertising can be substantiated, and whether there is a reasonable basis for the claims made in the advertising. In the past, FTC has issued enforcement policy statements stating that it will defer to FDA regarding the enforcement of certain kinds of food advertising, e.g., the use of health claims and nutrient content claims. FTC has not yet elaborated on how it plans to enforce advertising regarding rDNA biotechnologyderived foods, so it is not clear whether and to what extent FTC might follow any FDA policy that is issued with respect to rDNA biotechnology-derived foods.

Labeling Case Studies

In evaluating the labeling framework for rDNA biotechnology-derived foods, consideration should be given to at least three analogous situations: irradiated foods, milk from rBSTtreated cows, and organic foods.

Food Irradiation: Irradiation is defined by statute as a food additive, the only process that is so defined. It entails the treatment of a food with an FDA-approved energy source that kills bacteria or pests, prevents sprouting of root vegetables, or extends shelf life in some foods. Irradiation is an example of a process that triggers a label disclosure requirement because FDA determined that irradiation can render food materially different organoleptically, e.g., taste, smell, and texture.

Therefore, FDA determined by regulation that the fact that a food is irradiated is material, justifying the labeling requirement of a logo and a phrase such as "treated with irradiation". Despite some limited studies indicating good consumer acceptance of irradiated food, food processors generally took a conservative position in adoption of the technology. They concluded that irradiated products with the mandatory labeling would be avoided by consumers and could result in loss of sales, bad publicity, and loss of investment.

Other factors that may have inhibited use of irradiation are opposition by some activist groups, low-volume demand, overall cost of operation, high capital investment, technical expertise needed by workers, limited availability of suitable packaging, slow equipment development, and large sums of money already invested in alternative technologies. Nonetheless, there is considerable evidence that the irradiation labeling requirement slowed the food industry's adoption of this technology. Today, there are several recent examples where consumers have preferred an irradiated product to the traditional nonirradiated product.

One example is strawberries, where irradiation extends the shelf life of the raw fruit. Recent concerns about microbiological safety of foods have drawn the public's attention to the potential benefits of irradiation processing. As a result, some food processors are again considering further utilization of this technology. FDA determined that the process of irradiation caused food to differ significantly from its conventional counterpart, thus making irradiation of food a material fact that must be disclosed. The irradiation label disclosure requirements have been cited as at least one significant factor inhibiting the use of this pathogen-reducing technology.

Milk from rBST-Treated Cows: In the early 1990s, FDA approved treatment of dairy cows with recombinant bovine somatotropin (rBST), an rDNA biotechnology-derived version of a naturally occurring hormone that increases a cow's milk production. FDA determined that milk produced by cows treated with rBST was not significantly different from conventional milk. Nonetheless, significant controversy accompanied the introduction of rBST into the marketplace.

Some manufacturers attempted to address consumer interest in avoiding milk from rBST-treated cows by labeling milk products as "rBST-free." FDA discouraged "rBST-free" claims because they implied that there is some com- positional difference, such as the presence of rBST, between milk from treated and untreated cows. Rather, FDA encouraged the use of claims that address the production procedure rather than the product. So FDA announced that an appropriate way to phrase such an acceptable claim would be, "from cows not treated with rBST," as long as the statement also provided a context that did not imply a difference between the milks.

Organic Foods: The term "organic" has been used to describe foods grown without certain modern farming practices that some consumers find objectionable. The organic food movement began using statements concerning the production of foods without the use of certain types of commercial pesticides and fertilizers. The focus of the organic movement has expanded and centered on the societal goals of some citizens, including a reduction in the usage of agricultural chemicals, a healthier environment, more humane treatment of animals, greater worker safety, and enhanced food safety. The movement established production criteria that not only pertained to conditions for growing crops but also for labeling and distribution of such foods.

The organic movement originally enlisted several state governments to recognize or adopt documentation and inspection programs designed to demonstrate compliance with these criteria. In some cases, it has become necessary to provide separate production and distribution systems for organic and non-organic foods. To date, scientific evidence does not demonstrate that organic foods have superior nutritional or food safety benefits over non-organic foods. Therefore, FDA has deemed some claims on organic foods misleading when the term "organic" has been used in a manner that implied that the organic food is somehow superior to a similar non-organic food.

The regulations are intended to further establish a market for a niche category of "organic" foods desired by consumers. Under the proposed rule, every farm or other organic operation would have to develop and carry out an "organic plan" that would be approved and certified by a USDA-accredited agent. The NOP would include a "National List" that sets forth which chemical substances are permitted for use in organic production. The NOP also would create three categories of permissible label claims, each with its own criteria: "100 percent organic"; "organic"; and "made with organic."

Products labeled "100 percent organic" would have to be all organic product; products bearing the "organic" label would have to contain not less than 95% organically produced product; and products labeled "made with organic (specified ingredients)" would have to contain at least 50% organic ingredients. Any of the three label claims could be used, in accordance with requirements set forth in the regulations; e.g., all would have to bear the seal or logo of the certifying agent, anywhere on the package and on any other labeling or market information about the product.

With the exception of products labeled "100 percent organic," the listing for each organic ingredient would have to be qualified with the term "organic" in the ingredients statement. Products labeled "made with organic (specified ingredients)" would be subject to labeling limitations (e.g., maximum type size, no more than three organic ingredients may be listed), and, unlike products labeled "100 percent organic" or "organic," would not be allowed to bear the USDA Organic Seal.

In order for products containing less than 50% organic ingredients to use the term "organic," the label would have to declare the total percentage of organic ingredients on the information panel (the label panel that typically includes nutrition information, the ingredients statement, and similar information) and qualify each organic ingredient with the term "organic" in the ingredients statement.

The product would not be allowed to use "organic" anywhere else on the label or to bear the USDA Organic Seal or the seal or logo of any certifying agent. The proposed NOP rule is clear that rDNA biotechnology-derived and irradiated foods are not considered "organic." Any product made with what the proposed rule terms "exclud- ed methods" could not be labeled as "organic." USDA made this decision based on "overwhelming public opposition" to the use of rDNA biotechnology in organic production systems, even though the agency admitted "there is no current scientific evidence that use of excluded methods presents unacceptable risks to the environment or human health".

Thus, it may be possible for consumers wishing to avoid rDNA biotechnology- derived foods to purchase foods bearing one of the three "organic" label claims on the principal display panel. With organic products a rapidly growing percentage of the market-organic food sales in the U.S. have risen dramatically, from $78 million in 1980 to an estimated $6 billion in 2000, and projected annual growth is approximately 20%-organic foods appear to be a readily available option for consumers who wish to avoid rDNA biotechnology-derived foods.

The labeling of organic foods is an example of a voluntary program that focuses on production differences that are of significant consumer interest, even though they do not render foods materially different from their conventional counterparts. Under the constitutional restrictions described above, such distinctions may not be addressed through government-mandated disclosures, but may be freely described through voluntary label statements. To avoid confusion regarding the meaning of terms and to clarify rules in a manner that helps organic food processors and marketers avoid making misleading claims, Congress actively monitored USDA's development of standards. Improved clarity of labeling terms and greater efficiency associated with higher product volumes appear to be facilitating growth in organic foods.

FDA has not established special labeling requirements for foods derived using rDNA biotechnology. Yet, the general framework of food labeling regulation provides a series of food labeling requirements for rDNA biotechnology-derived foods.

Mandatory Disclosures: Constitutional restrictions specified in Central Hudson and the FFDCA prohibition of labeling that is misleading by virtue of omission of a material fact are important factors regarding mandatory label disclosures. Labeling requirements that apply to foods in general also apply to foods derived using rDNA biotechnology. To avoid a misleading presentation of the food, the label must reveal all material facts. In developing its labeling policy for rDNA biotechnology-derived foods, FDA considered public comments and scientific evidence regarding the presence of material facts about such foods. rDNA biotechnology- derived foods do not differ materially as a class of food from conventional foods.

Individual rDNA biotechnology-derived foods may or may not be significantly different from their conventional counterparts. FDA requires labeling of specific rDNA biotechnology-derived foods that differ significantly in composition, nutritional value, or safety from their conventional counterparts. Thus, if a food derived using rDNA biotechnology differs from its conventional counterpart such that the common or usual name no longer adequately describes the new food, the name must be changed or qualified to describe the difference.

If a safety or usage issue exists for the new food, a statement must be made on the label to describe the issue. For example, if a food derived using rDNA biotechnology has significantly different nutritional properties, its name must reflect the difference. Likewise, if a new food includes an allergen that consumers would not expect based on the name of the food, the presence of that allergen must be stated on the label. Some have advocated that the mandatory labeling requirements reach beyond disclosure of material facts regarding the food.

They have urged a blanket requirement for disclosure when a food is derived using rDNA biotechnology. In developing its 1992 labeling policy, FDA considered public comments and all available scientific evidence in connection with a possible blanket rDNA biotechnology disclosure requirement. FDA rejected such a blanket requirement because it was "not aware of any information showing that rDNA biotechnology- derived foods] differ from other foods in any meaningful or uniform way, or present any different or greater safety concern than foods developed by traditional plant breeding."

In the absence of a material fact to distinguish an rDNA biotechnologyderived food from its conventional counterpart, the imposition of a blanket disclosure requirement would be constitutionally suspect. As with the Vermont rBST label disclosure requirement that was ruled unconstitutional in International Dairy Foods v. Amestoy, the absence of a distinguishing characteristic of the rDNA biotechnology- derived food requires the government to demonstrate a substantial government interest to justify a label disclosure requirement. The Amestoy court specifically rejected "consumer curiosity" as basis for a substantial government interest.

Voluntary Claims: Required declarations only constitute a portion of the label information that serves the consumer's right to know. In a competitive marketplace, there are powerful incentives for the introduction of factors for distinguishing products that would be appreciated by significant population segments. When government action to compel labeling is involved, the distinguishing factors must be material facts regarding the food. Countless other factors may be advanced through voluntary label claims and, if appreciated by consumers, rewarded in the marketplace.

So foods that are not rDNA biotechnology-derived may be labeled as such in a truthful and non-misleading manner. Consumers who appreciate that distinction are served by such labeling. Government restrictions on misleading labeling can also influence the nature and even the availability of such voluntary claims. Naturally, the prohibition against false and misleading claims is an important service to the consumer's right to know. However, ambiguity in what may be viewed as false or misleading label statements may have a chilling effect on the marketing of such claims.

For example, a processor asserting that a product includes no rDNA biotechnology- derived ingredients must be able to substantiate that claim to provide reasonable assurance of its accuracy. Before a processor undertakes that risk, it may reasonably seek guidance as to how FDA expects such substantiation to be accomplished and what degree of purity is required to justify a claim that a product is free of rDNA biotechnology-derived ingredients. As detailed below, the answers to these questions may have significant economic effects that can greatly influence the availability of such claims.

The degree of purity is a complex issue that must be resolved with careful consideration of what such a claim is likely to mean to the consumer. For example, fat-free claims are permitted on products that have up to one-half gram of fat per serving because it is not feasible to measure lesser amounts of fat in foods. Moreover, the general public health objective is to limit fat in the diet, not eliminate it altogether. Further, the degree of purity a claim represents would be considered by FDA in determining appropriate methods of substantiation.

For example, a claim like "These ingredients were not genetically modified," which addressed the process by which the food was produced, would not imply the same degree of purity as a "Free of GM ingredients" claim. Supplier certifications are generally regarded as less accurate and less expensive than product testing. FDA would seek accuracy, but would likely be reluctant to require an excessively expensive substantiation method, since its costs could discourage use of the label statement.

A claim about the production process would likely be easier to verify than a claim about the composition of the food. Moreover, the potential for a claim about the absence of rDNA biotechnology- derived ingredients to be interpreted as misleading because it inaccurately implies superiority of the food over its counterpart may also discourage such claims. One need only consider the rBST experience to see that a simple "GM-free" claim may be regarded as misleading because it implies superiority in safety or environmental effect. FDA recommended the use of terminology that disclosed the use or non-use of rBST on the cow, rather than the presence or absence of rBST in the milk.

Therefore, as was sought by the organic foods movement, clear guidance as to government's expectations regarding voluntary claims can eliminate the regulatory uncertainty that may discourage investment in a new product line of foods that are to be marketed as free of rDNA biotechnology- derived ingredients. However, establishing a clear definition of "GMfree" through legislation or a regulatory mechanism will be a difficult process that would require general agreement from most stakeholders.

FDA announced that it plans to develop a guidance document regarding the voluntary labeling of food about the presence or absence of rDNA biotechnology-derived ingredients. FDA stated that it would use focus groups and seek public comment on its draft guidance. The Labeling Panel is mindful that not all voluntary claims regarding rDNA biotechnology-derived foods are claims that the food is free of rDNA biotechnology-derived ingredients. One of the first rDNA biotechnology- derived foods introduced to the market was the FlavrSavr tomato, which bore statements in labeling indicating that the tomato was developed through rDNA biotechnology. these claims also must be truthful and non-misleading.

International Policies

Because of the world market for food and food ingredients, U.S. labeling requirements and policies should not be examined in isolation. Rather, the subject should be considered in the context of international requirements and policies. In view of the growing conditions where cross-pollination might occur, the complexities of the harvesting, storage, and distribution practices, and other possibilities for inadvertent contamination of non-rDNA biotechnology- derived crops with rDNA biotechnology- derived crops, it would be necessary for any mandatory labeling program to include a level of inadvertent rDNA biotechnology-derived crop contamination above which labeling would be required.

This level should be consistent with the objectives of the labeling program and be acceptable to the program's proponents. It should also be legally defensible and practical to enforce, including scientifically appropriate, practical, and affordable analytical testing. There is currently no international consensus on what such a regulatory level should be. The regulation of rDNA biotechnology- derived foods differs widely in other countries. Some countries do not allow them to be imported at all, on the basis that not enough is known about the long-term effects of consuming rDNA biotechnology-derived foods.

Other countries permit such foods, with requirements that each food disclose on the label that it was produced using rDNA biotechnology. Such policies require foods at each stage of production, from raw agricultural product to finished consumer package, to bear a statement like "Contains GMO." Still other countries, like the U.S., compare the new plant variety to varieties produced using conventional breeding to identify differences for safety evaluation and to determine whether the differences need to be described on the food label.

These countries do not require statements that the food was derived using rDNA biotechnology. International debates on labeling of rDNA biotechnology-derived foods take place in a variety of forums and often are part of the discussion of the safety of the foods. Countries that require declaration of the fact that the food was derived using rDNA biotechnology tend to have less confidence in the safety of the food. This lower confidence level is not always based on scientific assessments that indicate higher potential risk. The consumer's right to know is an important factor in the rationale of those supporting mandatory labeling, particularly when consumer confidence in the regulatory system is low.

The primary international forum for discussion of labeling of rDNA biotechnology-derived foods is the Codex Alimentarius

Commission. Codex implements the International Food Standards Program under the joint sponsorship of the Food and Agriculture Organization (FAO) of the United Nations and the World Health Organization (WHO). Its purpose is to harmonize international food requirements to foster public health and international trade. However, not all participating countries adopt all Codex standards.

For several years, the Codex Committee on Food Labeling (CCFL) has been discussing how rDNA biotechnology- derived foods should be labeled, with the goal of having a single approach to labeling requirements. Two basic approaches to labeling are being considered by the CCFL. One, based on the principle of comparing new varieties of foods to those produced by conventional means, would require the description of any differences on the label.

That is, where the rDNA biotechnology-derived food differs significantly from the conventional version of the food in nutritional value, in physical or handling properties, or by the presence of an allergen not indicated by the name of the food, this difference would have to be disclosed on the label. The other approach to labeling involves the conclusion that the use of rDNA biotechnology, either in the food or to produce the food, is itself a difference that requires disclosure.

The U.S., Canada, and several other countries support the approach that only the significant differences need to be disclosed, while the European countries and others favor what is called the "mandatory labeling" approach. A number of countries are currently developing national policies, and several of them are likely to require mandatory labeling. Even though the discussion is ongoing in Codex and will likely continue for several more years, the effects of differing national rDNA biotechnology labeling policies are already being felt.

For example, European countries are major markets for U.S. grains and food ingredients. Some European retailers have decided to exclude rDNA biotechnology-derived ingredients from foods manufactured for their own "store brands". Some U.S. producers have decided to return to growing non-rDNA biotechnology-derived crops rather than lose this market. Some European food manufacturers, particularly infant formula firms, have said they will not accept rDNA biotechnology-derived ingredients.

Some U.S. manufacturers of processed corn and soybean consumer products have announced that they will not accept any rDNAderived biotechnology crops for processing. While Codex is the major forum for discussion of international food labeling requirements, another organization has also developed a position on labeling. The protocol has been interpreted to provide that "living modified organisms" (LMO) intended for "food, feed, or processing" must be identified as LMO. While the U.S. signed the Cartagena Protocol, it has not been ratified by the U.S. Senate.

As a result, it has no legal effect in the U.S. at this time. As interpreted by the U.S. Department of State, the Protocol requires international shipments of bulk LMO commodities to be accompanied by documentation stating "May contain LMOs"; however, the Protocol does not impose consumer product labeling requirements. The European Union (EU), Japan, South Korea, Australia, and New Zealand have all passed, or are considering, laws requiring that food containing rDNA biotechnology-derived ingredients be labeled.

These countries are in various stages of implementing such requirements. The EU has been considering this issue for a substantial period of time and has approved a de minimis threshold level of 1%. This de minimis threshold would apply separately to each ingredient used in the product, and only apply to those situations where the presence of the rDNA biotechnology-derived material was unintentional.

The two most common methods for detecting rDNA biotechnology-derived materials are the polymerase chain reaction (PCR)-based methods, which detect genetically modified

DNA sequences, and immunoassays, which measure levels of proteins expressed by inserted DNA sequences. Most PCR-based assays are qualitative in nature and are routinely used to determine if rDNA is present in a sample. PCR is a laboratory-based technique requiring trained staff and specialized equipment. It is extremely sensitive and capable of detecting one or a few copies of a gene.

DNA extraction methods need to be optimized for each food matrix, as various food components inhibit the reagents used in the assay. In general, DNA is not detectable in highly heat-treated foods, hydrolyzed plant proteins, purified lecithin, starch derivatives, and refined oils derived from rDNA biotechnology- derived crops. The use of Real- Time PCR, which uses fluorescence to monitor the PRC amplification process, shows significant promise in resolving this problem.

However, the equipment is, at present, very expensive, ranging in price from $36,000 to $95,000. Because these methods are extremely sensitive, there is a significant risk of cross-contamination resulting in false positives. The Enzyme Linked Immunosorbent Assay (ELISA) uses one antibody to bind the specific protein and an antibody conjugated to an enzyme whose product generates a color that can be easily visualized and quantified.

Nonquantitative immunoassays are also available in the form of plate or lateral strip formats. Immunoassays are less sensitive than PCR and therefore are less susceptible than PCR to false positives caused by minor levels of contamination. Assay validation is important because of the large diversity of food matrices. In addition, it is important to determine sequence homology with other proteins that might be present in the sample.

Per-sample costs are in the range of $2 to $10, although up-front costs for assay development and generation of antibodies and protein standards are high. Results are available within minutes. Immunoassays are not capable of distinguishing between different rDNA biotechnology- derived events that express similar protein characteristics. Since proteins are denatured by many food processing methods, immunoassays are best used for raw commodities or minimally processed ingredients.

There are some common problems no matter which assay method is used. Currently, there are no internationally recognized sampling methods or agreement on the number or size of samples. Sampling plans must be scientifically and statistically sound and take into account the potential heterogeneity of samples due to adventitious contamination from cross-pollination or during distribution. Positive and negative reference standards are not readily available to validate analytical methods or to assess the performance of methods and laboratories.

As new rDNA biotechnology-derived crops are developed, sequence information must be shared by the technology generators so that new assays can be developed. Ultimately, hundreds of different assays may be necessary. Several agencies in Europe and the U.S., including the European Commission's Joint Research Centre, the U.S. National Institute of Standards and Technology, and USDA's Grain Inspection, Packers, and Stockyards Administration, are working on reference standards and validation programs for rDNA testing methods. Validation and standardization of sampling and testing methods are essential to resolving disputes regarding the status of food ingredients, and for regulatory authorities responsible for enforcing mandatory labeling laws.

Market Segmentation of Crops

Labeling of rDNA biotechnologyderived foods, whether mandatory or voluntary, must be premised on discrete distribution channels for the underlying commodities, such as soybeans and corn. The majority of processed foods contain ingredients derived from corn and/or soybeans. One major trade association has estimated that 70% of processed foods contain corn or soyderived ingredients.

In fact, corn and soy serve as the source of thousands of ingredients used in processed foods. Common ingredients derived from corn include corn oil, corn starch, corn flour, corn meal, maltodextrins, corn syrup, and dextrose. Soy-derived ingredients include soy oil, bran, flour, sauce and meal, soy protein isolates and concentrates, texturized vegetable protein, lecithin, and mono- and diglycerides. Many processed foods contain multiple corn and soy-derived ingredients. For example, a typical cake mix contains hydrogenated soybean oil, modified corn starch, mono- and diglycerides, dextrose, and soy lecithin.

Some corn and soy-derived ingredients serve as secondary ingredients (e.g, carriers for flavors, colors, or vitamins) and may not be listed on the label. For many of these ingredients, substitutes are not available that provide the same functionality, texture, and taste. Since corn and soy are managed as commodity ingredients and labeling is not required in the U.S., many of the processed foods on the market today likely contain ingredi- ents derived from rDNA biotechnology-derived crops. The world markets for corn and soybeans are currently undergoing a process of segmentation in response to concerns in the EU and elsewhere.

It is clear that consumer concerns about rDNA biotechnology in some countries will determine the marketability of rDNA biotechnology-derived corn and soybeans. Several major food processors in different countries have announced that they will only accept non-rDNA biotechnology-derived crops. Thus, some supplies in the world market must be guaranteed rDNA biotechnology-free to meet this emerging demand. Two major processors announced that they would pay a premium for a particular non-rDNA biotechnology-derived soybean product that has been bred to resist one particular herbicide.

In addition, they announced they would reject any rDNA biotechnology-derived corn not accepted in EU markets during the 1999 growing year. In 1999, this program provided an 18-cent-per-bushel premium (approximately 3.5%) for soybeans. A similar approach is expected for the 2000 crop year. One major supplier of rDNA biotechnology- derived seeds announced that it would help growers of its rDNA biotechnology-derived crops find domestic market outlets for varieties that are not approved by the EU.

Its suppliers segregate rDNA biotechnology- derived and non-rDNA biotechnology- derived crops. This segmentation goes beyond rejection of rDNA biotechnology-derived products not approved by the EU, and was said to be a response to growing consumer requests for such segmentation. Some U.S. firms paid a premium for non-rDNA biotechnology-derived products, but the size of the premium and the extent of demand for non-rDNA biotechnologyderived products was very uncertain. In addition to changes in demand, another source of uncertainty is the regulatory status of rDNA biotechnologyderived crops in other major producing countries, such as Brazil.

Although Brazil does not allow rDNA biotechnology-derived crops, a significant portion of the crops produced are genetically modified because seed is smuggled into the country. Whether segmentation by country source or segmentation within market channels can take place will be an important determinant of eventual marketing costs. For example, EU buyers are able to source their corn imports in 2000 from countries other than the U.S., but they must still buy a significant share of their soybean product imports from the U.S. Thus, there is greater market incentive to develop guaranteed non-rDNA biotechnologyderived soybean channels in the U.S.

As of spring 2000 there had been relatively little market differentiation of rDNA biotechnology-derived crops in the U.S. The marketing costs and price premiums associated with the development of a segmented market include any premium paid to a producer to supply a particular variety, the costs of segregation in storage and handling, and the costs of verifying that the crop is truly not rDNA biotechnology-derived.

The complexity of the U.S. grain-marketing channels makes it difficult and expensive to segregate crops when variety is the only clear difference, as testing for genetic modification is currently an expensive alternative. Whether the objective is to label or to ban rDNA biotechnology-derived crops, only three alternative marketing strategies exist: test the product at selected points in the market channel; accept producer assurances at the first handler and maintain identity through the market channel; or use third-party supervision and certification from seed to final processing. Each alternative has its advantages and its limitations.

Final Product Testing: This approach provides assurance that the sample selected for testing meets the standards of the buyer. It focuses on the attributes of the product rather than on the process by which it was produced and delivered. The disadvantages are the costs and uncertainties of sampling and testing. Current sampling methods provide a low level of confidence that a large bulk shipment is adequately represented by the sample analyzed. The standard error for even the best sampling strategy (e.g., automatic diverter samplers in the inbound or outbound grain stream) is large for the current low tolerance levels.

Sampling of inbound deliveries by farmers presents the same problem of obtaining representative samples, and it has the additional problem of time required for testing and segregating, given the speed with which inbound vehicles must be unloaded. A single test that can be applied at the first-handler level for testing for the entire range of possible genetic modifications has not yet been developed. As discussed above, the most common rDNA biotechnology test method currently used is PCR. It is a very sensitive test which is most useful in detecting the presence of rDNA biotechnology-derived materials, but it takes 3-5 days to complete and costs several hundred dollars per sample.

The ELISA approach takes 5-20 minutes and costs less than $10 per sample. However, neither approach has been accepted as being quantitatively reliable because of a lack of standardized sampling techniques and reliable control standards.

For labeling purposes, what to test for is also important. It is possible, at least with PCR, to test for the promoter or marker DNA, which is common to many rDNA biotechnology-derived organisms, or to test for the specific genes that confer the desirable traits. The first type of test would identify only whether a crop was rDNA biotechnology-derived; the second type would identify what type of modification had taken place.

Producer Validation and Market Segregation: This strategy segregates the non-rDNA biotechnology-derived crops at the beginning of the market channel. If the product is shipped in containers dedicated to non-rDNA biotechnology-derived grain, guaranteeing the process will also guarantee the final product. However, there are problems with this approach in addition to the cost of segregation. Although producers may know the variety and the extent of potential crosspollination at the time of harvest, much commercial grain is delivered by commercial haulers who do not have this information. Without prior contracts or arrangements with the producer, producer validation and certifications have questionable reliability.

This strategy also reqûires grain handling establishments to maintain separate facilities, as it is not feasible to clean facilities of all rDNA biotechnology-derived grain between loads. Some establishments designate one of their facilities for handling nonrDNA biotechnology-derived grain, thus simplifying the problem of identification at the time of delivery. Yet a major obstacle to maintaining purity through the rest of the market channel still exists, because trucks, rail cars, barges, and port equipment must also use dedicated equipment to guarantee that all rDNA biotechnology-derived grain has been excluded.

Third-Party Certification: This strategy reduces the danger of misinformation, questionable methods of isolation in the field, and incomplete knowledge on the part of the producer or distributor. The strategy is well known and frequently used for delivering food-quality corn and soybeans to foreign destinations. Organic, pesticide-free, and variety- specific qualifications are common in international trade, but third-party certification adds significant costs per bushel.

Illinois grain handlers currently use third-party certification from seed to river elevator and shipping in containers or small-volume segregated barge loads that are transferred directly from barge to vessel to avoid contamination. This strategy is based on the premise that it is more effective to guarantee the process than to guarantee the product.

No shipper can guarantee that a very small amount of rDNA biotechnology-derived product will not be introduced into a shipment from any of many sources, including the storage facility in the destination country. Instead of guaranteeing 100% purity, this approach provides assurances that the grain has been handled in such a way as to minimize the possibility of contamination.

Cost of Market Segregation: The marketing costs associated with specialty grains in Illinois. Such specialty grains have particular characteristics, such as oil or protein content, that bring high value in particular end-use markets. An average additional handling cost of $0.17 per bushel for corn and $0.48 per bushel for soybeans in 1998, over and above the premium for specialty characteristics. Internetbased e-markets to contract with farmers for non-rDNA biotechnologyderived corn or soybeans showed that the market is still sorting out methods of verification and premiums for non-rDNA biotechnology-derived crops.

The current means of verification for non-rDNA biotechnology-derived products included all of the possibilities discussed above. The firms interviewed used spot testing, segregated on-farm storage, segregated onsite storage at the elevator, and segregated transportation measures. Some elevators do not do any testing and relied on the word of the farmer regarding the non-rDNA biotechnology-derived product. Widely varying premiums for non-rDNA biotechnology-derived product that was contracted for in the previous spring.

It is useful to think about rDNA biotechnology-derived and non-rDNA biotechnology-derived corn and soybeans as separate products. There is substitution in supply, but in some countries there is little substitution in demand. rDNA biotechnology-derived varieties reduce costs of production and/or increase yields for some U.S. producers. If rDNA biotechnology-derived crops have lower costs, producers presumably will look to be compensated for producing non-rDNA biotechnology-derived crops at higher cost for the EU.

The following scenario seems possible: Some portion of EU and other demand will be for guaranteed nonrDNA biotechnology-derived crops. At the same time, there will be widespread adoption of rDNA biotechnology- derived varieties in major producing countries. The demand for nonrDNA biotechnology-derived varieties will be met from segmented market channels that will develop in all exporting countries, and this supply will carry a marketing premium. In addition to this marketing premium, producers will receive a price premium to cover the higher costs of production of non-rDNA biotechnology-derived crops.

However, this premium will be in relation to the somewhat lower world prices for corn and soybeans due to increased total supplies brought about by adoption of rDNA biotechnology. The costs of the producer premium and marketing premium are likely to be passed on to buyers in the EU and elsewhere. The long-run costs of this segmentation are more difficult to predict. These long-run costs arise from the disincentives to invest in rDNA biotechnology that result from lack of consumer acceptance.

Depending on the degree of purity demanded by any market segmentation, the current grain handling and distribution system may need to be modified, or a new, dedicated system for non-rDNA biotechnology-derived products developed. For example, while the practicality of currently available test methods for detecting rDNA biotechnology-derived organisms could be improved, the sensitivity of these methods greatly exceeds the capabilities of existing U.S. grain handling and distribution systems to deliver non-rDNA biotechnology-derived crops.

In other words, it may be unrealistic to expect a grain handling and distribution system that tolerates, for example, 1% corn as "foreign material" in soybeans to exclude a comparable, if not lower, level of rDNA biotechnology- derived soybeans-which would not even be considered "foreign material"-in a lot of non-rDNA biotechnology- derived soybeans.

Consumer Reactions

Most U.S. consumers are not concerned about the safety of foods derived using rDNA biotechnology. Eighty percent of consumers are confident that food in the supermarket is safe. Most consumers support application of rDNA biotechnology in agriculture and food production. A majority (53%) believe that rDNA biotechnology does not pose a health hazard. A much smaller number (27%) believe that rDNA biotechnology does pose a serious health hazard, while 20% have no opinion. The majority of consumers believe that rDNA biotechnology will provide benefits for them or their families within five years.

Labels are a valuable source of information for consumers. People indicate that they consult labels to obtain accurate information as to product ingredients and nutritional content. Information on labels and in government publications generate greater trust than any other information source. It is well recognized that the method of asking a question can influence consumer response. Some surveys indicate that the majority of consumers want foods derived by rDNA biotechnology to be labeled; however, consumers frequently respond affirmatively when asked if they want additional information.

When asked if labels should contain a variety of information, 85% of consumers desired disclosure that rDNA biotechnology was used. Fewer consumers expressed interest in labeling if rDNA biotechnology was used to produce ingredients or processing aids, such as chymosin for cheese making. Other information, such as use of pesticides or country of production, was desired by 94% and 80%, respectively. Consumers were not told that labeling would entail additional cost. Focus group research indicated that most consumers believed that cost would be minimal.

Chapter 8

Recommended International Code of Practice - General Principles of Food Hygiene

The General Principles of Food Hygiene follow the food chain from primary production through to final consumption, highlighting the key controls at each stage. In brief, they give guidance on the design and facilities of premises, in-process control, required support programmes of sanitation and personal hygiene and consideration of hygiene controls once the product has left the production premises. They recommend an HACCP-based approach wherever possible to enhance food safety, as described in the Hazard Analysis and Critical Control Point (HACCP) system and guidelines for its application.

The controls described in the General Principles of Food Hygiene are internationally recognized as essential to ensuring the safety and suitability of food for consumption. The general principles are recommended to governments, industry, including individual primary producers, manufacturers, processors, food service operators and retailers, and consumers alike. All have a responsibility in ensuring safe food for the consumer and reducing the incidence of foodborne illness and food spoilage.

New challenges facing the food industry include new food production, preparation and distribution techniques, changing eating habits and increasing volumes of food being transported around the world. Furthermore, opportunities for international trade are enhanced where food is produced in a strictly hygienic environment, and a country that follows strict hygienic practices gains a reputation as a producer of safe food.

CODEX ALIMENTARIUS COMMISSION

In Latin, meaning of the Codex Alimentarius is "Food Code" or "Food Law". To familiarize the participants with the role and activities of the Codex Alimentarius Commission and the importance of the Codex codes of hygienic practice, standards, guidelines and other recommendations consequent to the Final Act of the Uruguay Round of the General Agreement on Tariffs and Trade (GATT) and in particular the Agreement on the Application of Sanitary and Phytosanitary Measures (SPS Agreement) and the Agreement on Technical Barriers to Trade (TBT Agreement), which recognize Codex standards, guidelines and other recommendations as the specifically identified baseline for consumer protection.

Suggested methods of instruction

— Lecture

— Video presentation

Aids

— Overhead transparencies/slides

— Handout: list of final Codex texts

— Codex Alimentarius video

Reference

— This is Codex Alimentarius

Time frame

— One hour

Content

— Introduction to the Codex Alimentarius Commission
— The Uruguay Round agreements

The Codex Alimentarius is a collection of food standards, codes of practice and other recommendations presented in a uniform way. Codex standards, guidelines and other recommendations ensure that food products are not harmful to the consumer and can be traded safely between countries. Food safety standards are defined in the Agreement on the Application of Sanitary and Phytosanitary Measures as those relating to food additives, veterinary drug and pesticide residues, contaminants, methods of analysis and sampling, and codes and guidelines of hygienic practice. Codex food safety standards are to be used as the reference point for the World Trade Organization in this area. There are more than 300 Codex standards, guidelines and other recommendations relating to food quality composition and safety.

The Codex Alimentarius has resulted in evaluations of the safety of over 760 food additives and contaminants and the setting of more than 2 500 maximum limits for pesticide residues and more than 150 for veterinary drug residues. In addition, CAC has established guideline levels for a number of environmental and industrial contaminants in foods. Food hygiene has been a major area of activity of CAC since the commission's establishment. The Codex Committee on Food Hygiene, hosted by the Government of the United States, was founded in 1963. Because food hygiene is best regulated at the production and processing stage in the exporting country, the committee's main outputs have been codes of hygienic practice rather than end-product microbiological standards.

Taking this philosophy a step further, CAC has adopted the Guidelines for the application of the Hazard Analysis Critical Control Point (HACCP) system through its Committee on Food Hygiene. In doing this, it has recognized HACCP as a tool to assess hazards and establish control systems that focus on preventive measures instead of relying primarily on end-product testing. CAC has been actively revising much of its work in recent years to stress the so-called horizontal aspects of food regulation, including food hygiene. New considerations such as risk analysis and the determination of equivalence in different food control systems have an impact on the new approach to international food hygiene regulations.

URUGUAY ROUND AGREEMENTS

The Uruguay Round negotiations were the first to deal with the liberalization of trade in agricultural products, an area excluded from previous rounds of negotiations. The Uruguay Round also included negotiations on reducing non-tariff barriers to international trade in agricultural products and concluded with two binding agreements: the Agreement on the Application of Sanitary and Phytosanitary Measures (SPS Agreement) and the Agreement on Technical Barriers to Trade (TBT Agreement). The agreements will be applied by members of WTO, and the general terms are also applicable to countries that are not WTO members.

Application of SPS Measures Agreements

The Sanitary and Phytosanitary Measures (SPS) Agreement confirms the right of WTO member countries to apply measures necessary to protect human, animal and plant life and health. This right was included in the original 1947 GATT as a general exclusion from the other provisions of the agreement provided that "such measures are not applied in a manner which would constitute a means of arbitrary or unjustifiable discrimination between countries where the same conditions prevail, or a disguised restriction on international trade".

Despite this general condition for the application of national measures to protect human, animal and plant life and health, it had become apparent that national sanitary and phytosanitary measures had become, whether by design or accident, effective trade barriers. The SPS Agreement therefore sets new rules in an area previously excluded from GATT disciplines. The purpose of the SPS Agreement is to ensure that measures established by governments to protect human, animal and plant life and health, in the agricultural sector only, are consistent with obligations prohibiting arbitrary or unjustifiable discrimination on trade between countries where the same conditions prevail and are not disguised restrictions on international trade.

It requires that, with regard to food safety measures, WTO members base their national measures on international standards, guidelines and other recommendations adopted by the FAO/WHO Codex Alimentarius Commission where they exist. This does not prevent a member country from adopting stricter measures if there is a scientific justification for doing so, or if the level of protection afforded by the Codex standard is inconsistent with the level of protection generally applied and deemed appropriate by the country concerned. The SPS Agreement covers all food hygiene measures and food safety measures such as the control of residues of veterinary drugs, pesticides and other chemicals used in meat production. In addition, it covers animal and plant quarantine measures.

The SPS Agreement states that any measures taken that conform to international Codex standards, guidelines or other recommendations are deemed to be appropriate, necessary and non-discriminatory. Furthermore, the SPS Agreement calls for a programme of harmonization of national requirements based on international standards. This work is guided by the WTO Committee on Sanitary and Phytosanitary Measures, to which representatives of CAC, the International Office of Epizootics (OIE) and the International Plant Protection Convention (IPPC) are invited.

Technical Barriers to Trade Agreement

TBT Agreement is a revision of the agreement of the same name first developed under the Tokyo Round of GATT negotiations in the 1970s. Examples given in the TBT Agreement of legitimate TBT measures are those with the objective of national security or the prevention of deceptive practices. The objective of the agreement is to prevent the use of national or regional technical requirements, or standards in general, as unjustified technical barriers to trade.

The agreement covers standards relating to all types of products including industrial and agricultural products, with the exception of aspects of food standards related to sanitary and phytosanitary measures. It includes numerous measures designed to protect consumers against deception and economic fraud. Examples of food standards covered by the TBT Agreement are those related to quality and labelling. The TBT Agreement basically provides that all technical standards and regulations must have a legitimate purpose and that the impact or cost of implementing the standard must be proportional to the purpose of the standard. It also says that if there are two or more ways of achieving the same objective, the least trade-restrictive alternative should be followed.

The agreement also places emphasis on international standards, WTO members being obliged to use international standards or parts of them except where the international standard would be ineffective or inappropriate in the national situation. The TBT Agreement does not include a programme of harmonizing national standards.

Codex standards, guidelines and other recommendations take on unprecedented importance with respect to consumer protection and international food trade. As a result, the work of the Codex Alimentarius Commission - including the Guidelines for the application of the Hazard Analysis Critical Control Point (HACCP) system - has become the reference for international food safety requirements. In view of this it is imperative that the Codex guidelines for the application of HACCP be unequivocal in their guidance; otherwise conflicts on food safety grounds may arise.

GENERAL PRINCIPLES OF FOOD HYGIENE

The Codex General Principles of Food Hygiene lay a firm foundation for ensuring food hygiene. They follow the food chain from primary production through to the final consumer, highlighting the key hygiene controls at each stage, and recommend an HACCP-based approach whenever possible to enhance food safety as described in the Hazard Analysis and Critical Control Point (HACCP) system and guidelines for its application. These controls are internationally recognized as essential to ensure the safety and suitability of food consumption.

The general principles are recommended to governments, industry and consumers alike. The requirements of the General Principles of Food Hygiene are considered to be the foundation for the development of an HACCP-based system for ensuring food safety. The application of the General Principles of Food Hygiene and of good manufacturing practices (GMPs) allows the producer to operate within environmental conditions favourable to the production of safe food. In implementing an HACCP system in an establishment, the first step is to review existing programmes for compliance with the General Principles of Food Hygiene and GMPs and to verify whether all the necessary controls and documentation are in place. The importance of these programmes cannot be overstated, as they are the foundation of the implementation of the HACCP plan.

Inadequate programmes may lead to additional critical control points that would have to be identified, monitored and maintained under the HACCP plan. In summary, adherence to the General Principles of Food Hygiene and GMPs will simplify the implementation of HACCP plans and will ensure that the integrity of HACCP plans is maintained and that the manufactured product is safe. In order to harmonize or standardize the approach, the training is formatted around the Recommended International Code of Practice - General Principles of Food Hygiene. The application of HACCP principles should be preceded by compliance with the Principles of Food Hygiene and appropriate Codex commodity codes of practice. These controls are internationally recognized as necessary to ensure the safety and suitability of food for consumption.

The term "this document" whenever used in the general principles refers to the Recommended International Code of Practice-General Principles of Food Hygiene, and any allusion to the Annex refers to the annex to that document. The term "contamination" in the General Principles of Food Hygiene refers to the contamination of food by microbial pathogens, chemicals, foreign bodies, spoilage agents, objectionable taints and unwanted or diseased matter, e.g. sawdust or decomposed material. The general principles also use the terms "food safety" and "suitability for consumption". The former is used in the context of ensuring that food does not cause illness or injury to consumers, the latter in the context of distinguishing if food is spoiled or otherwise not suitable for normal human consumption.

CODEX GENERAL PRINCIPLES OF FOOD HYGIENE

To introduce the participants to the Codex General Principles of Food Hygiene as a firm foundation for ensuring food safety and to their prerequisite relationship to the development of effective HACCP or equivalent systems

Suggested method of instruction

— Lecture

Aids

— Overhead transparencies/slides

— Handout

Reference

— Recommended International Code of Practice - General Principles of Food Hygiene [CAC/RCP 1-1969, Rev. 3 (1997)], Sections I and II - reproduced below in shaded boxes

Time frame

— One hour

Content

— What are the General Principles of Food Hygiene?

— Objectives

— Scope, use and definitions

— Structure

Codex General Principles of Food Hygiene are given below:

— identify the essential principles of food hygiene applicable throughout the food chain, including primary production through to the final consumer, to achieve the goal of ensuring that food is. safe and suitable for human consumption;

— recommend an HACCP-based approach as a means to enhance: food safety;

— indicate how to implement those principles; and

— provide a guidance for specific codes which may be needed for - sectors of the food chain; processes; or commodities; to amplify the hygiene requirements specific to those areas.

Food chain: This document follows the food chain from primary production to the final consumer, setting out the necessary hygiene conditions for producing food which is safe and suitable for consumption. The document provides a base-line structure for other, more specific, codes applicable to particular sectors. Such specific codes and guidelines should be read in conjunction with this document and Hazard Analysis and Critical Control Point (HACCP) system and guidelines for its application.

Roles of governments, industry and consumers: Governments can consider the contents of this document and decide how best they should encourage the implementation of these general principles to:

— protect consumers adequately from illness or injury caused by food; policies need to consider the vulnerability of the population, or of different groups within the population;

— provide assurance that food is suitable for human consumption;

— maintain confidence in internationally traded food; and,

— provide health education programmes which effectively communicate the principles of food hygiene to industry and consumers.

Industry should apply the hygienic practices set out in this document to:

— provide food which is safe and suitable for consumption;

— ensure that consumers have clear and easily understood information, by way of labelling and other appropriate means, to enable them to protect their food from contamination and growth/survival of foodborne pathogens by storing, handling and preparing it correctly; and

— maintain confidence in internationally traded food.

*Definitions:*For the purpose of this Code, the following expressions have the meaning stated:

— *Cleaning:* the removal of soil, food residue, dirt, grease or other objectionable matter.

— *Contaminant*: any biological or chemical agent, foreign matter, or other substances not intentionally added to food which may compromise food safety or suitability.

— *Contamination*: the introduction or occurrence of a contaminant in food or food environment.

— *Disinfection:* the reduction, by means of chemical agents and/or physical methods, of the number of microorganisms in the environment, to a level that does not compromise food safety or suitability.

— *Establishment*: any building or area in which food is handled and the surroundings under the control of the same management.

— *Food hygiene:* all conditions and measures necessary to ensure the safety and suitability of food at all stages of the food chain.

— *Hazard*: A biological, chemical or physical agent, in, or condition of, food with the potential to cause an adverse health effect.

— *HACCP*: A system which identifies, evaluates, and controls hazards which are significant for food safety.

— *Food handler:* any person who directly handles packaged or unpackaged food, food equipment and utensils, or food contact surfaces and is therefore expected to comply with food hygiene requirements.

— *Food safety*: assurance that food will not cause harm to the consumer when it is prepared and/or eaten according to its intended use.

— *Food suitability*: assurance that food is acceptable for human consumption according to its intended use.

— *Primary production*: those steps in the food chain up to and including, for example, harvesting, slaughter, milking, fishing.

Structure: The Codex Alimentarius General Principles of Food Hygiene include an Introduction, an Annex [Hazard Analysis and Critical Control Point (HACCP) system and guidelines for its application] and ten sections as follows:

— Section I - Objectives of the general principles of food hygiene

— Section II - Scope and use of the document

— Section III - Primary production

— Section IV - Establishment: design and facilities

— Section V - Control of operation
— Section VI - Establishment: maintenance and sanitation
— Section VII - Establishment: personal hygiene
— Section VIII-Transportation
— Section IX - Product information and consumer awareness
— Section X - Training

Sections III to X are addressed individually in the remaining modules (3 to 10) of this section of the manual.

PRIMARY PRODUCTION

To introduce the trainees to the importance of identifying potential food safety hazards at the primary production stage of the food chain and the necessity of controlling or minimizing hazards at this stage in order to reduce the likelihood of introducing a hazard that may adversely affect the safety of food, or its suitability for consumption, at later stages of the food chain; to review the role of government and the importance of the Codex Alimentarius with respect to the control of pesticide residues, residues of veterinary drugs and contaminants in foods

Primary production should be managed in a way that ensures that food is safe and suitable for its intended use. Where necessary, this will include:

— avoiding the use of areas where the environment poses a threat to the safety of food;
— controlling contaminants, pests and diseases of animals and plants in such a way as not to pose a threat to food safety;
— adopting practices and measures to ensure food is produced under appropriately hygienic conditions.

To reduce the likelihood of introducing a hazard which may adversely affect the safety of food, or its suitability for consumption, at later stages of the food chain. The major challenge in encouraging the management of primary production is the integration of the role of government with the role of the primary producer. There is a need for governments to participate in the control of hazards associated with primary production through the regulation of pesticides and veterinary drugs, the identification and control of environmental hazards and the development of "good practices" documents. Education and training programmes at the primary production level should be developed to facilitate the management of primary production.

Environmental Hygiene: Potential sources of contamination from the environment should be considered. In particular, primary food production should not be carried on in areas where the presence of potentially harmful substances would lead to an unacceptable level of such substances in food.

Hygienic production of food sources: The potential effects of, primary production activities on the safety and suitability of food should be considered at all times. In particular this includes identifying any specific points in such activities where a high probability of contamination may exist and taking specific measures to minimize that probability. The HACCP-based approach may assist in the taking of such measures.

Producers should as far as practicable implement measures to:

— control contamination from air, soil, water, feedstuffs, fertilizers, pesticides, veterinary drugs or any other agent used in primary production;

— control plant and animal health so that it does not pose a threat to human health through food, consumption, or adversely affect the suitability of the product; and

— protect food sources from faecal and other contamination.

In particular, care should be taken to manage wastes, and store harmful substances appropriately. On-farm programmes which achieve specific food safety goals are becoming an important pact of primary production and should be encouraged.

Handling, Storage and Transport: Procedures should be in place to:

— sort food and food ingredients to segregate material which is evidently unfit for human consumption;

— dispose of any rejected material in a hygienic manner; and

— protect food and food ingredients from contamination by pests, or by chemical, physical or microbiological contaminants or other objectionable substances during handling, storage and transport.

Care should be taken to prevent, so far as reasonably practicable, deterioration and spoilage through appropriate measures which may include controlling temperature, humidity, and/or other controls. Cleaning, maintenance and personnel hygiene at primary production. Appropriate facilities and procedures should be in place to ensure that:

— any necessary cleaning and maintenance is carried out effectively; and

— an appropriate degree of personal hygiene is maintained.

Role of Government

Governments need to provide guidance to primary producers and to establish regulatory control programmes to ensure food safety and wholesomeness at the primary production level. Hazards associated with primary production may or may not be eliminated or reduced to acceptable levels depending on the subsequent processing and handling of the primary food products. Possible health risk may occur because of primary products excessively contaminated with microorganisms or toxins that could affect the health of consumers. An understanding of how pathogens are introduced during primary production is essential in the development of appropriate interventions and effective control mechanisms.

In many cases, however, primary production control measures have not yet been defined that will provide a means to control certain hazards. More research is needed to establish the ecology of pathogenic microorganisms so that appropriate intervention strategies can be devised for the reduction of pathogens at the beginning of the food chain. Other programmes, such as good husbandry practices, could be used to reduce the incidence of pathogens, thereby reducing risk at later stages of the food chain. Simple changes such as minimizing the amount of moisture, mud and faeces buildup on animal hide, hair, feathers or skin would be expected to reduce microbial levels. This type of approach would be particularly important prior to shipment for slaughter.

A number of other factors, such as reduction of animal stress and reduction of feed and water contamination, when combined with other practices would result in overall pathogen reduction and minimization of risk. Similar control strategies may be developed for plant food sources. Interventions might include avoidance of fertilization with manures containing viable pathogens and avoidance of sewage-contaminated growing areas. Another type of health risk can occur if primary products are contaminated as a result of improper use of pesticides or veterinary drugs or because of environmental contamination.

The improper use of pesticides or veterinary drugs in primary production can result in unsafe residues of these substances in the food. In addition to providing the health protection benefits, effective pesticide and veterinary drug use and residue control programmes and environmental control programmes enable a country to participate in international food trade with greater confidence; an effective residue control programme can serve as the foundation for certifying the safety of the country's exported food products, as well as providing assurance of the safety of food products imported into the country.

Scientific evaluation of the safety of pesticides or veterinary drugs and the acceptable levels for human consumption is a long and rigorous task which may not need to be performed in each country, especially among developing countries. In such evaluations the interested country could use the technical expertise of international organizations such as the Joint FAO/WHO Expert Committee on Food Additives or the Codex Committee on Pesticide Residues. The elements in establishing an effective national programme for the control of pesticide or veterinary drug residues in food should include but not necessarily be limited to the following:

— Establishing the regulatory authority/authorities for implementing inspection programmes and laboratory analysis
— Establishing an integrated inspection programme, including a residue control programme for the inspection of foods
— Compiling registers of veterinary drugs and pesticides used in the country, including products manufactured in the country and products imported into the country
— Elaborating regulations concerning the distribution of veterinary drugs and pesticides, providing procedures for authorized sale, manufacture, import, distribution and use of such products
— Elaborating procedures for determining the safety and efficacy of veterinary drugs and pesticide residues.
— Establishing procedures for monitoring through sampling of food products for pesticide residues and veterinary drugs
— Selecting the methods of analysis to be used for pesticide and veterinary drug residues
— Implementing a laboratory quality assurance programme to ensure the highest quality of analytical results
— Developing educational programmes for primary producers and veterinarians, instructing on the proper use of pesticides and veterinary drugs and encouraging the use of preventive measures to reduce the occurrence of residues in food

HACCP plans may not need to be developed on an individual basis for primary producers but may be developed by experts and recommended to primary producers as "good practice recommendations". Education and training programmes should be relied upon to introduce practices that, in effect, may represent a change in the manner in which farms and other primary food production operations are managed.

To introduce the participants to Section IV of the Codex General Principles of Food Hygiene and to examine the importance and requirements of good hygienic design and construction, including appropriate siting, design and construction of premises, equipment and facilities to control risks of contamination. Depending on the nature of the operations, and the risks associated with them, premises, equipment and facilities should be located, designed and constructed to ensure that:

— contamination is minimized;

— design and layout permit appropriate maintenance, cleaning and disinfections and minimize airborne contamination;

— surfaces and materials, in particular those in contact with food, are non-toxic in intended use and, where necessary, suitably durable, and easy to maintain and clean;

— where appropriate, suitable facilities are available for temperature, humidity and other controls; and

— there is effective protection against pest access and harbourage.

Potential sources of contamination need to be considered when deciding where to locate food establishments, as well as the effectiveness of any reasonable measures that might be taken to protect food. Establishments should not be located anywhere where, after considering such protective measures, it is clear that there will remain a threat to food safety or suitability. In particular, establishments should normally be located away from:

— environmentally polluted areas and industrial activities which pose a serious threat of contaminating food;

— areas subject to flooding unless sufficient safeguards are provided;

— areas prone to infestations of pests;

— areas where wastes, either solid or liquid, cannot be removed effectively.

Equipment should be sited so that it:

— permits adequate maintenance and cleaning;

— functions in accordance with its intended use; and

— facilitates good hygiene practices, including monitoring.

Design and layout: Where appropriate, the internal design and layout of food establishments should permit good food hygiene practices, including protection against cross-contamination between and during operations by foodstuffs. The following should be considered in providing protection against cross-contamination:

— Activities should be adequately separated by physical or other effective means where cross-contamination may result.

— Buildings and facilities should be designed to facilitate hygienic operations by means of a regulated flow in the process from the arrival of the raw material at the premises to the finished product. Where appropriate, blueprints and/or process flow diagrams should be available.

Internal Structures and Fittings: Structures within food establishments should be soundly built of durable materials and be easy to maintain, clean and where appropriate, able to be disinfected. In particular the following specific conditions should be satisfied where necessary to protect the safety and suitability of food:

— the surfaces of walls, partitions and floors should be made of impervious materials with no toxic effect in intended use;

— walls and partitions should have a smooth surface up to a

height appropriate to the operation;

— floors should be constructed to at low adequate drainage and cleaning;

— ceilings and overhead fixtures should be constructed and finished to minimize the build up of dirt and condensation, and the shedding of particles;

— windows should be easy to clean, be constructed to minimize the build up of dirt and where necessary, be fitted with removable and cleanable insect-proof screens. Where necessary, windows should be fixed;

— doors should have smooth, non-absorbent surfaces, and be easy to clean and, where necessary, disinfect;

— working surfaces that come into direct contact with food should be in sound condition, durable and easy to clean, maintain and disinfect. They should be made of smooth, non-absorbent materials, and inert to the food, to detergents and disinfectants under normal operating conditions.

Premises and structures covered here include market stalls, mobile sales and street vending vehicles, temporary premises in which food is handled such as tents and marquees. Such premises and structures should be sited, designed and constructed to avoid, as far as reasonably practicable, contaminating food and harbouring pests. In applying these specific conditions and requirements, any food hygiene hazards associated with such facilities should be adequately controlled to ensure the safety and suitability of food.

Equipments and containers coming into contact with food, should be designed and constructed to ensure that, where necessary, they can be || adequately cleaned, disinfected and maintained to avoid the contamination of food. Equipment and containers should be made of materials with no toxic effect in intended use. Where necessary, equipment should be durable and movable or capable of being disassembled to allow for maintenance, cleaning, disinfection, monitoring and, for example, to facilitate inspection for pests. The manufacturer should have an effective written preventive maintenance programme to ensure that equipment that may affect food is maintained in proper working order. This should include:

— A list of equipment requiring regular maintenance

— The procedures and frequencies of maintenance are based on the equipment manufacturer's manual or equivalent or on operating conditions that could affect the condition of the equipment

The preventive maintenance programme should be adhered to. Equipment should be maintained to ensure the absence of any physical or chemical hazard potentials, e.g. inappropriate repairs, flaking paint and rust, excessive lubrication.

Food control and monitoring equipment: In addition to the general requirements: equipment used to cook, heat treat, cool, store or freeze food should be designed to achieve the required food temperatures as rapidly as necessary in the interests of food safety and suitability, and maintain them effectively. Such equipment should also be designed to allow temperatures to be monitored and controlled. Where necessary, such equipment should have effective means of controlling and monitoring humidify, air-flow and any other characteristic likely to have a detrimental effect on the safety or suitability of food. These requirements are intended to ensure that:

— harmful or undesirable microorganisms or their toxins are eliminated or reduced to safe levels or their survival and growth are effectively controlled;

— where appropriate, critical limits established in HACCP-based plans can be monitored; and

— temperatures and other conditions necessary to food safety and suitability can be rapidly achieved and maintained.

Containers for waste, by-products and inedible or dangerous

substances, should be specifically identifiable, suitably constructed and, where appropriate, made of impervious material. Containers used to hold dangerous substances should be identified and, where appropriate, be lockable to prevent malicious or accidental contamination of food.

Facilities

Water supply: Art adequate supply of potable water with appropriate facilities for its storage, distribution and temperature control, should be available whenever necessary to ensure the safety and suitability of food. Potable water should be as specified in the latest edition of WHO guidelines for drinking water quality, or water of a higher standard. Non-potable water shall halve a separate system. Non-potable water systems shall be identified and shall not connect with, or allow reflux into, potable water systems.

Drainage and waste disposal: Adequate drainage and waste disposal systems and facilities should be provided. They should be designed and constructed so that the risk of contaminating food or the potable water supply is avoided.

Cleaning: Adequate facilities, suitably designated, should be provided for cleaning food, utensils and equipment. Such facilities should have an adequate supply of hot and cold potable water where appropriate.

Personnel hygiene facilities and toilets: Personnel hygiene facilities should be available to ensure that an appropriate degree of personal hygiene can be maintained and to avoid contaminating food. Where appropriate, facilities should include:

- adequate means of hygienically washing and drying hands, including wash basins and a supply:! of hot and cold (or suitably temperature controlled) water;
- lavatories of appropriate hygienic design; and
- adequate changing facilities for personnel.

Such facilities should be suitably located and designated.

Temperature control: Depending on the nature of the food operations undertaken, adequate facilities should be available for heating, cooling, cooking, refrigerating and freezing food, for storing refrigerated or frozen foods, monitoring food temperatures, and when necessary, controlling ambient temperatures to ensure the safety and suitability of food. Adequate means of natural or mechanical ventilation should be provided, in particular to:

— minimize air-borne contamination of food, for example, from aerosols and condensation droplets;

— control ambient temperatures; and

— control odours which might affect the suitability of food; and

— control humidity, where necessary, to ensure the safety and suitability of food.

Ventilation systems should be designed and constructed so that air does not flow from contaminated areas to clean areas and, where necessary, they can be adequately maintained and cleaned.

Adequate natural or artificial lighting should be provided to enable the undertaking to operate in a hygienic manner. Where necessary, lighting should not be such that the resulting colour is misleading. The intensity should be adequate to the nature of the operation. Lighting fixtures should, where appropriate, be protected to ensure that food is not contaminated by breakages.

Where necessary, adequate facilities for the storage of food, ingredients and non-food chemicals (e.g. cleaning materials, lubricants, fuels) should be provided. Where appropriate, food storage facilities should be designed and constructed to:

— permit adequate maintenance and cleaning;

— avoid pest access and harbourage;

— enable food to be effectively protected from contamination during storage; and

— where necessary, provide an environment which minimizes the deterioration of food.

The type of storage facilities required will depend on the nature of the food. Where necessary, separate, secure storage facilities for cleaning materials and hazardous substances should be provided. Some points are to be considered

— Ingredients requiring refrigeration should be stored at 4°C or less and should be appropriately monitored. Frozen ingredients should be stored at temperatures that do not permit thawing.

— Ingredients and packaging materials should be handled and stored in such a manner as to prevent damage and/or contamination.

— Rotation of ingredients and where appropriate packaging materials should be controlled to prevent deterioration and spoilage.

— Humidity-sensitive ingredients and packaging materials should be stored under appropriate conditions to prevent deterioration.

— Non-food chemicals should be received and stored in a dry/ well-ventilated area.

— Non-food chemicals should be stored in designated areas such that there is no possibility for cross-contamination of food or food contact surfaces.

— Where required for ongoing use in food handling areas, these chemicals should be stored in a manner that prevents contamination of food/food contact surfaces or packaging materials.

— Chemicals should be stored and mixed in clean, correctly labelled containers.

— Chemicals should be dispensed and handled only by authorized and properly trained personnel.

— Finished product should be stored and handled under conditions that prevent deterioration.

— Stock rotation should be controlled to prevent deterioration that could present a health hazard.

— Returned defective or suspect product should be clearly identified and isolated in a designated area for appropriate disposition.

— Finished product should be stored and handled in a manner to prevent damage. For example, stacking heights should be controlled and forklift damage avoided.

CONTROL OF OPERATION

To introduce the trainees of the Codex General Principles of Food Hygiene and to examine the considerations regarding the key aspects of control systems, incoming material requirements, packaging, water, management, documentation and records and recall systems.

To produce food which is safe and suitable for human consumption by:

— formulating design requirements with respect to raw materials, composition, processing, distribution, and consumer use to be met in the manufacture and handling of specific food items; and

— designing, implementing, monitoring and reviewing effective control systems.

Food Hazards Control

Food business operators should control food hazards through the use of systems such as HACCP. They should:

— identify any steps in their operations which are critical to the safety of food;

— implement effective control procedures at those steps;

— monitor control procedures to ensure their continuing effectiveness; and

— review control procedures periodically, and whenever the operations change.

These systems should be applied throughout the food chain to control food hygiene throughout the shelf-life of the product through proper product and process design.

Control procedures may be simple, such as checking stock rotation, calibrating equipment, or correctly loading refrigerated display units. In some cases a system based on expert advice, and involving documentation, may be appropriate. A model of such a food safety system is described in Hazard Analysis and Critical Control Point (HACCP) system and guidelines for its application.

General Control Procedures Examples

Product formulae: Current written formulae provide a basis for assessment of food additives, nutritional requirements, food allergens and the scheduled process. The processor should consider the following:

— Written master formulae should be available.

— The formulae should contain all details of the formulation, including identification of specific ingredients and additives (e.g. concentration, type) and amounts of additives and ingredients.

— The master formulae should be current for the products being processed.

Food additives: Inadequate control of food additives could result in chemical or biological hazards.

— The manufacturer should ensure that all food additives used are permitted for use in the particular food and meet all requirements of the food law.

— The manufacturer should have specifications for all food additives.

— The manufacturer should require that all food additives have the identity and purity required to meet food grade quality.

— The manufacturer should obtain certification/verification from the supplier that each lot of food additives meets the requirements of the food law, e.g. specification sheets and clear identification of the grade on the additive package.

— The manufacturer should verify and demonstrate through calculations that food additives are used within the maximum level specified in the food law.

The manufacturer should have controls in place during preparation or blending to ensure that food additives are permitted and are used within allowable levels. These include:

— Clear identification of additives

— Accurate measurement

— Adequate blending for homogeneity

Nutritional requirements: The manufacturer should have control over the product formulation to ensure that all nutritional requirements and claims are met. Formulation controls are necessary to prevent hazards that could result from excesses, inadequacies and omissions of nutrients, for example in the case of special dietary foods, infant formulae, meal replacements, fortified foods and foods for which nutritional claims have been made.

— Nutrients used should be permitted in accordance with the food law.

— The nutrient content of the product should be accurately reflected on the label.

— The manufacturer should have specifications for nutrients.

— The manufacturer should receive from the supplier a certificate of analysis accompanying each lot of nutrient;

for nutrients used in foods that are the sole source of nutrition, each certificate should be verified through analysis.

— The manufacturer should verify and demonstrate through calculations that nutrients are used within the limits specified in the food law.

The manufacturer should have controls in place for nutrient addition during preparation or blending to ensure that the levels comply with regulatory and label requirements. These controls should include:

— Clear identification of nutrient

— Proper storage and handling to maintain nutrient stability

— Accurate measurement

— Adequate blending for homogeneity

Label accuracy: The manufacturer should have procedures in place to ensure that label information accurately represents the composition and formulation of the product. Controls are necessary to prevent the presence of undeclared ingredients or misinformation concerning product composition. The manufacturer should ensure that the label information provides the public with accurate information related to net contents; manufacturer's, packer's and/or distributor's names; and instructions for proper handling and preparation at home. The following controls should be included:

— New label review

— Incoming label review for accuracy/correctness

— Formulation changes/substitutions

Allergens: The manufacturer should have controls in place to prevent the presence of undeclared allergens in the product. Allergens are those ingredients that will elicit an allergic response in sensitive individuals. Areas that may require control include:

— Misdirection of ingredients

— Rework

— Contamination by undeclared ingredients

— Ingredient carryover

— Ingredient substitutions

— Carryover from equipment, e.g. from product changeovers

Product preparation/blending: Critical factors specified in the formulation should be controlled during preparation and blending to prevent physical, chemical, nutritional and biological hazards. Inadequate control of critical factors associated with product preparation or blending could result in underprocessing, formation of toxins, presence of undeclared allergens, violative levels of food additives or nutritional hazards. In thermal processing, the manufacturer should have controls for critical factors identified in the validated process. Examples of critical factors are:

— Size, e.g. dicing, grinding, slicing

— Temperature treatment, e.g. heating, blanching, defrosting, cooling.

— Moisture, e.g. rehydration, concentration (viscosity, brix)

— Proportioning, e.g. weight, volume (weighing, volumetric control, metering)

— pH/acidity (pH measurement, titratable acidity)

For microbial control, the manufacturer should control time and temperature during preparation, blending, and holding of in-process materials to prevent conditions that could result in excessive microbial growth or in the production of enterotoxin by Staphylococcus aureus.

Hygiene Control System Aspects

Time and Temperature Control

Inadequate food temperature control is one of the most common causes of foodborne illness or food spoilage. Such controls include time and temperature of cooking, cooling, processing and storage. Systems should be in place to ensure that temperature is controlled effectively where it is critical to the safety and suitability of food. Temperature control systems should take into account:

— the nature of the food, e.g. its water activity, pH, and likely initial level arid types of microorganisms;
— the intended shelf-life of the product;
— the method of packaging and processing; and!-how the product is intended to be used, e.g. further cooking/ processing or ready-to-eat.

Such systems should also specify tolerable limits for time and temperature variations. Temperature recording devices should be checked at regular intervals and tested for accuracy. Other steps which contribute to food hygiene may include, for example:

— chilling
— thermal processing
— irradiation
— drying
— chemical preservation
— vacuum or modified atmospheric packaging

Microbiological, Physical and chemical contamination

Management systems offer an effective way of ensuring the safety and suitability of food. Where microbiological, chemical or physical specifications are used in any food control system, such specifications should be based on sound scientific principles and state, where appropriate, monitoring procedures, analytical methods and action limits.

Pathogens can be transferred from one food to another, either by direct contact or by food handlers, contact surfaces or the air. Raw, unprocessed food should be effectively separated, either physically or by time, from ready-to-eat foods, with effective intermediate cleaning and, where appropriate, disinfection, Access to processing areas may need to be restricted or controlled. Where risks are particularly high, access to processing areas should be only via a changing facility. Personnel may need to be required to put on clean protective clothing including footwear and wash their hands before entering.

Surfaces, utensils, equipment, fixtures and fittings should be thoroughly cleaned and where necessary disinfected after raw food, particularly meat and poultry, has been handled or processed. Systems should be in place to prevent contamination of foods by foreign bodies such as glass or metal shards from machinery, dust, harmful fumes and unwanted chemicals. In manufacturing and processing, suitable detection or screening devices should be used where necessary.

Incoming Material Requirements

Raw materials or ingredients should, where appropriate, be inspected and sorted before processing. Where necessary, laboratory tests should be made to establish fitness for use. Only sound, suitable raw materials or ingredients should be used. Stocks of raw materials and ingredients should be subject to effective stock rotation. Prevention of health hazards begins with control of incoming materials. Inadequate incoming ingredient controls could result in product contamination and/or underprocessing. The degree of control exercised over incoming ingredients is appropriate to the risk. The manufacturer should control incoming ingredients through one of the following programmes or equivalent.

The first three options apply to ingredients that may be critical factors where further processing is not likely to eliminate a hazard. The fourth option applies to ingredients that are not likely to have an impact on the safety of the food. Periodic evaluation of incoming ingredients:

— The manufacturer should have written specifications for ingredients.

— Purchasing specifications should include a provision for compliance with the food law.

— The manufacturer should maintain a documented history of adherence to specifications for each supplier, e.g. analytical results.

— The manufacturer should obtain a certificate of analysis for each lot (code).

— A statistically representative sample should be taken to verify the accuracy of the certificates of analysis at a scheduled frequency.

— A new history of adherence to specifications should be established when a firm changes suppliers, purchases ingredients from a new supplier or purchases a new ingredient from an existing supplier or when spot checks do not agree with the certificate of analysis. One hundred percent lots inspected:

— The manufacturer should have written specifications for ingredients.

— The manufacturer should obtain a certificate of analysis for each lot (code).

— Each incoming lot should be sampled according to a predetermined sampling plan and analysed for adherence to specifications.

Vendor certification: When the manufacturer relies on vendor certification the following minimum requirements should be in place:

— The manufacturer should have written specifications for ingredients.

— The manufacturer should have documentation to demonstrate adequate knowledge of the vendor's process, e.g. process flow charts, on-site evaluations, identification of critical control points, specifications, control limits, monitoring programmes and frequencies, corrective action and verification procedures.

— The manufacturer should have data to demonstrate the capability of the vendor's process to manufacture consistently within specifications.

— Prior to implementation of a periodic monitoring programme the firm should analyse an appropriate number of consecutive lots to establish a historical database and confirm adherence to specifications.

— The manufacturer should conduct periodic monitoring to verify adherence to specifications.

— The manufacturer should conduct vendor audits to validate the status of the vendor certification programme.

Where incoming ingredients are not likely to impact on the safety of the food:

— The manufacturer should have written specifications for these ingredients.

— Purchasing specifications should include a provision for compliance with the food law.

— The supplier should provide assurance that the ingredients meet specifications.

When ingredients are found not to meet specifications, the manufacturer investigates and identifies the root cause. If the ingredients do not meet specifications but have not been used, the case is not considered a deviation; however, if it is possible that ingredients that do not meet specifications have been used, the manufacturer should initiate deviation/correction control procedures.

Packaging

Packaging design and materials should provide adequate

protection for products to minimize contamination, prevent damage, and accommodate proper labelling. Packaging materials or gases where used must be non-toxic and not pose a threat to the safety and suitability of food under the specified conditions of storage and use. Where appropriate, reusable packaging should be suitably durable, easy to clean and, where necessary; disinfect. Where appropriate, the manufacturer should control incoming packaging materials from the supplier using an approach similar to the options suggested for incoming ingredients.

Water

Only potable water should be used In food handling and processing, with the following exceptions:

— for steam production, fire control and other similar purposes not connected with food; and

— in certain food processes, e.g. chilling, and in food handling areas, provided this does not constitute a hazard to the safety and suitability of food.

Water recirculated for reuse should be treated and maintained in such a condition that no risk to the safety and suitability of food results from its use. The treatment process should be effectively monitored. Recirculated water which has received no further treatment and water recovered from processing of food by evaporation or drying may be used, provided its use does not constitute a risk to the safety and suitability of food.

Ice and steam: Ice should be made from water. Ice and steam should be produced, handled and stored to protect them from contamination. Steam used in direct contact with food or food contact surfaces should not constitute a threat to the safety and suitability of food.

— Ice used as an ingredient or in direct contact with food should be made from potable water and protected from contamination.

— Boiler treatment chemicals used should be acceptable to the regulator and should not result in unacceptable residues.
— Boiler feed water should be tested regularly, and its chemical treatment should be controlled to prevent contamination.
— The steam supply should be generated from potable water and should be adequate to meet operational requirements.
— Traps should be provided as necessary to ensure adequate condensate removal and elimination of foreign materials.

Management and Supervision

The type of control and supervision needed will depend on the size of the business, the nature of its activities and the types of food involved. Managers and supervisors should have enough knowledge of food hygiene principles and practices to be able to judge potential risks, take appropriate preventive and corrective action, and ensure that effective monitoring and supervision takes place.

Records and Documentation

Where necessary, appropriate records of processing, production and distribution should be kept and retained for a period that exceeds the shelf-life of the product. Documentation can enhance the credibility and effectiveness of the food safety control system. The following are requirements for all record-keeping activities:

— Records should be legible and permanent and should accurately reflect the actual event, condition or activity.
— Errors or changes should be identified in a manner such that the original record is clear, e.g. struck out with a single stroke and initialed near the correction or change.
— Each entry on a record should be made by the responsible person at the time the event occurs. The completed records should be signed and dated by the responsible person.

— Critical records, e.g. records related to the adequacy of the thermal process and the achievement of a hermetic seal, should be signed and dated by a qualified individual designated by management prior to distribution of the product. All other records should be reviewed at an appropriate frequency to provide an early indication of potentially serious deficiencies.

— Records should be retained for one year after the expiry date on the label or container or, if there is no expiry date, for two years after the date of sale.

— Records should be maintained at the manufacturing plant and should be available upon request.

Recall Procedures

Where a product has been withdrawn because of an immediate health hazard, other products which are produced under similar conditions, and which may present a similar hazard to public; health, should be evaluated for safety and may need to be withdrawn. The need for public warnings should be considered. Recalled products should be held under supervision until they are destroyed, used for purposes other than human consumption, determined to be safe for human consumption, or reprocessed in a manner to ensure their safety. There should be a written procedure for recall, which should include:

— The person or persons responsible, e.g. recall coordinator(s)

— The roles and responsibilities for coordination and implementation of a recall

— Methods to identify, locate and control recalled product

— A requirement to investigate other products that may be affected by the hazard and that should be included in the recall

— A procedure for monitoring the effectiveness of the recall to the appropriate level of distribution specified in the recall notice

Recall information should include the following:

— Amount of product produced, in inventory and distributed

— Name, size, code or lot numbers of food recalled

— Area of distribution of product, e.g. local, national, international

— Reason for the recall

Product code identification

— Each pre-packaged food should have permanent, legible code marks or lot numbers and where required expiry or "best before" dates on the packages.

— Code marks used and the exact meaning of the codes should be available.

The manufacturer should be capable of producing accurate information on a timely basis to verify that all affected product can be rapidly identified and removed from the marketplace. This can be demonstrated by the manufacturer as follows:

— Records of customer names, addresses and telephone numbers available for the lot tested

— Records of production, inventory and distribution by lot available for the lot tested

— Periodic testing to verify the capability of the procedure for rapid identification and control of a code lot of potentially affected product and for reconciling the amount of product produced, in inventory and in distribution; identification and correction of any deficiencies in the recall procedure

Distribution records: Distribution records should contain sufficient information to permit traceability to a particular code or lot number. The following minimum information should be required for distribution records:

— Product identification and size

— Lot number or code
— Quantity
— Customers' names, addresses and telephone numbers to the initial level of product distribution

MAINTENANCE AND SANITATION

Establishments and equipment should be kept in an appropriate state of repair and condition to:
— facilitate all sanitation procedures;
— function as intended, particularly at critical steps;
— prevent contamination of food, e.g. from metal shards, flaking plaster, debris and chemicals.

Procedures and Methods of Cleaning

Cleaning should remove food residues and dirt which may be a source of contamination. The necessary cleaning methods and materials will depend on the nature of the food business. Disinfection may be necessary after cleaning. Cleaning chemicals should be handled and used carefully and in accordance with manufacturers' instructions and stored, where necessary, separated from food, in clearly identified containers to avoid the risk of contaminating food. Cleaning can be carried out by the separate or the combined use of physical methods, such as heat, scrubbing, turbulent flow, vacuum cleaning or other methods that avoid the use of water, and chemical methods using detergents, alkalis or acids. Cleaning procedures will involve, where appropriate:
— removing gross debris from surfaces;
— applying a detergent solution to loosen soil and bacterial film and hold them in solution or suspension;
— rinsing with water which: to remove loosened soil and residues of detergent;

- dry cleaning or other appropriate methods for removing and collecting residues and debris; and
- where necessary, disinfection.

Cleaning and disinfection programmes should ensure that all parts of the establishment are appropriately clean, and should include the cleaning of cleaning equipment. Cleaning and disinfection programmes should be continually and effectively monitored for their suitability and effectiveness and where necessary, documented. Where written cleaning programmes are used, they should specify:

- areas, items of equipment and utensils to be cleaned;
- responsibility for particular tasks;
- method and frequency of cleaning; and
- monitoring arrangements.

Where appropriate, programmes should be drawn up in consultation with relevant specialist expert advisors. The manufacturer should have a written cleaning and sanitation programme for all equipment which includes:

- The name of responsible person
- The frequency of the activity
- Chemicals and concentration used
- Temperature requirements
- Procedures for cleaning and sanitizing

The procedures for cleaning and sanitizing are different depending on whether the equipment is cleaned out of place (COP), e.g. hand-cleaned, or cleaned in place (CIP). For COP equipment, the procedures should be specified as follows:

- Identification of equipment and utensils
- Disassembly/reassembly instructions as required for cleaning and inspection

— Identification of areas on equipment requiring special attention
— Method of cleaning, sanitizing and rinsing

For CIP equipment, the procedures should be specified as follows:

— Identification of lines and/or equipment
— CIP set-up instructions
— Method of cleaning, sanitizing and rinsing
— Disassembly/reassembly instructions as required for cleaning and inspection

The manufacturer should have a written cleaning and disinfection programme for premises which specifies areas to be cleaned, method of cleaning, person responsible and frequency of the activity. Special sanitation and housekeeping procedures required during processing should be specified within the document, e.g. removal of product residues during breaks.

Pest Control Systems

Good hygiene practices should be employed to avoid creating an environment conducive to pests. Pests pose a major threat to the safety and suitability of food; Pest infestations can occur where there are breeding sites and a supply of food. Good sanitation, inspection of incoming materials and good monitoring can minimize the likelihood of infestation and thereby limit the need for pesticides.

Buildings should be kept in good repair and condition to prevent pest access and to eliminate potential breeding sites. Holes, drains and other places where pests are likely to gain access should be kept sealed. Wire mesh screens, for example on open windows, doors and ventilators, will reduce the problem of pest entry. Animals should, wherever possible, be excluded from the grounds of factories and food processing plants.

Harbourage and infestation: The availability of food and water encourages pest harbourage and infestation. Potential food sources should be stored in pest-proof containers and/or stacked above the ground and away from walls. Areas both inside and outside food premises should be kept clean. Where appropriate, refuse should be stored in covered, pest-proof containers. Establishments and surrounding areas should be regularly examined for evidence of infestation.

Eradication: Pest infestations should be dealt with immediately and without adversely affecting food safety or suitability. Treatment with chemical, physical or biological agents should be carried out without posing a threat to the safety or suitability of food. There should be an effective written pest control programme for the premises and equipment that includes:

— The name of the person and the operator assigned responsibility for pest control

— Where applicable, the name of the pest control company or the name of the person contracted for the pest control programme

— The list of chemicals used, the concentration, the location where applied, method and frequency of application

— A map of trap locations

— The type and frequency of inspection to verify the effectiveness of the programme

— Pesticides used should be acceptable to the food control regulatory authorities and should be used in accordance with the label instructions.

— Treatment of equipment, premises or ingredients for pest control should be conducted in such a manner as to ensure that the permitted maximum residue limit is not exceeded, e.g. by limiting the number of fumigation treatments per lot.

— Birds and animals, other than those intended for slaughter, should be excluded from establishments.

Waste Management

Waste must not be allowed to accumulate in food handling, food storage, and other working areas and the adjoining environment except so far as is unavoidable for the proper functioning of the business. Suitable provision must be made for the removal and storage of waste. Waste stores must be kept appropriately clean.

— Adequate facilities and equipment should be provided and maintained for the storage of waste and inedible material prior to removal from the establishment. These facilities should be designed to prevent contamination.

— Containers used for waste should be clearly identified, leakproof and, where appropriate, kept covered.

— Waste should be removed and containers cleaned and sanitized at an appropriate frequency to minimize contamination potential.

Sanitation systems should be monitored for effectiveness, periodically verified by means such as audit pre-operational inspections or, where appropriate, microbiological sampling of environment and food contact surfaces and regularly reviewed and adapted to reflect changed circumstances.

PERSONAL HYGIENE

To ensure that those who come directly or indirectly into contact with food are not likely to contaminate food by:

— maintaining an appropriate degree of personal cleanliness;

— behaving and operating in an appropriate manner.

Status of Health

People known, or suspected, to be suffering from, or to be a carrier of a disease or illness likely to be transmitted through food, should not be allowed to enter any food handling area if there is a likelihood of their contaminating food. Any person so affected should immediately report illness or symptoms of illness to the management. Medical examination of a food handler should be carried out if clinically or epidemiologically indicated.

— The manufacturer should have and enforce a policy to prevent personnel known to be suffering from or known to be carriers of a disease transmissible through food from working in food handling areas.

— The manufacturer should require that employees advise management when they are suffering from a communicable disease likely to be transmitted through food.

— Employees having open cuts or wounds should not handle food or food contact surfaces unless the injury is completely protected by a secure waterproof covering, e.g. rubber gloves.

Conditions which should be reported to management so that the need for medical examination and/of possible exclusion from food handling can be considered, include:

— jaundice

— diarrhoea

— vomiting

— fever

— sore throat with fever

— visibly infected skin lesions

— discharges from the ear, eye or nose;

Personal Cleanliness.

Food handlers should maintain a high degree of personal cleanliness and, where appropriate, wear suitable protective clothing, head covering and footwear. Cuts and wounds, where personnel are permitted to continue working, should be covered by suitable waterproof dressings. Personnel should always wash their hands when personal cleanliness may affect food safety, for example:

— at the start of food handling activities;

— immediately after using the toilet; and

— after handling raw food or any contaminated material, where this could result in contamination of other food items; they should avoid handling ready-to-eat food, where appropriate.

— All persons should wash their hands upon entering food handling areas, before starting work, after handling contaminated materials, after break and after using toilet facilities. Where necessary to minimize microbiological contamination, employees should use disinfectant hand dips.

— Protective clothing, hair covering, footwear and/or gloves appropriate to the operation that the employee is engaged in, e.g. effective hair coverings for employees in production areas, should be worn and maintained in a sanitary manner.

Personal Behaviour

People engaged in food handling activities should refrain from behaviour which could result in contamination of food, for example:

— smoking;

— spitting;

— chewing creating;

— sneezing or coughing over unprotected food.

Personal effects such as jewellery, watches, pins or other items should not be worn or brought into food handling areas if they pose a threat to the safety and suitability of food.

— Any behaviour that could result in contamination of food, such as eating, use of tobacco or chewing gum or unhygienic practices such as spitting, should be prohibited in food handling areas.

— All persons entering food handling areas should remove jewellery and other objects that could fall into or otherwise contaminate food. Jewellery that cannot be removed, such as wedding bands and medical alerts, should be covered.

— Personal effects and street clothing should not be kept in food handling areas and should be stored in such a manner as to prevent contamination.

Visitors

Visitors to food manufacturing, processing or handling areas should, where appropriate, wear protective clothing and adhere to the other personal hygiene provisions.

— Access of personnel and visitors should be controlled to prevent contamination. The traffic pattern of employees should not result in cross-contamination of the product.

TRANSPORTATION

Measures should be taken where necessary to:

— protect food from potential sources of contamination;

— protect food from damage likely to render the food unsuitable for consumption; and

— provide an environment which effectively controls the growth of pathogenic or spoilage microorganisms and the production of toxins in food.

Food must be adequately protected during transport. The type of conveyances or containers required depends on the nature of the food and the conditions under which it has to be transported. Where necessary, conveyances and bulk containers should be designed and constructed so that they:

— do not contaminate foods or packaging;
— can be effectively cleaned and, where necessary, disinfected;
— permit effective separation of different foods or foods from non-food items where necessary during transport;
— provide effective protection from contamination, including dust and fumes;
— can effectively maintain any temperature, humidity, atmosphere and other conditions necessary to protect food from harmful or undesirable microbial growth and deterioration likely to render it unsuitable for consumption; and
— allow any necessary temperature, humidity and other conditions to be checked.

Conveyances and containers for transporting food should be kept in an appropriate state of cleanliness, repair and condition. Where the same conveyance or container is used for transporting different foods, or non-foods, effective cleaning and, where necessary, disinfection should take place between loads. Where appropriate, particularly in bulk transport, containers and conveyances should be designated and marked for food use only and be used only for that purpose. The manufacturer should verify that carriers are suitable for the transportation of food. For example:

— Carriers should be inspected by the manufacturer on receipt and prior to loading to ensure they are free from contamination and suitable for the transportation of food.
— The manufacturer should have a programme in place to demonstrate the adequacy of cleaning and sanitizing. For example, for bulk carriers a written cleaning and sanitizing procedure should be available.
— Where the same carriers are used for food and non-food loads, procedures should be in place to restrict the type of non-food loads to those that do not pose a risk to subsequent

food loads after an acceptable cleaning or to food loads in the same shipment. For example/the manufacturer may require a cleaning certificate and a record of the previous material transported prior to loading or unloading of dual-use tankers, or may have a programme in place to verify the adequacy of cleaning, e.g. tanker inspection, sensory evaluation of ingredients and/or analysis, as appropriate.

— Carriers should be loaded, arranged and unloaded in a manner that prevents damage and contamination of the food.

— Bulk tanks should be designed and constructed to permit complete drainage and to prevent contamination.

— Where appropriate, materials used in carrier construction should be suitable for food contact.

Transportation and Distribution

In many cases, transportation firms may be handling a variety of products in addition to food, which adds to the complexity of the situation. The first step is to identify those circumstances that pose a significant health risk, such as improper handling of sensitive products or ineffective cleaning or sanitizing of transportation vehicles. For example, inadequate control of temperatures during transportation and distribution can contribute to microbial growth, formation of mycotoxins, spoilage and/or deterioration of certain products.

A recent example of improper practice has served to underscore needs in this area. An outbreak of salmonellosis in the United States was caused by the hauling of pasteurized ice-cream mix in a tanker truck that had previously carried raw eggs. This example illustrates the importance of preventing cross-contamination. HACCP plans developed by the food industry must consider the control of temperatures and contamination during the transport of foods. A food business operator may require an HACCP plan as a condition of doing business with a particular transportation firm. Properly designed HACCP-based good transportation practices for the transportation and distribution sector may be a more appropriate approach than HACCP plans.

General education programmes are needed to alert food transporters to the potential hazards that can be associated with the transportation and distribution of food products. Requirements for handling and distribution of food products or ingredients must be developed by food manufacturers, and these requirements must be communicated to transportation and distribution businesses. Transporters or storage facilities should be required to take proper hygienic measures to protect the food and should be required to keep and retain records that will document their adherence to food safety plans.

PRODUCT INFORMATION AND CONSUMER AWARENESS

Products should bear appropriate information to ensure that:

— adequate and accessible information is available to the next person in the food chain to enable them to handle, store, process, prepare and display the product safely and correctly;

— the lot or batch can be easily identified and recalled if necessary.

Consumers should have enough knowledge of food hygiene to enable them to:

— understand the importance of product information;

— make informed choices appropriate to the individual; and

— prevent contamination and growth or survival of foodborne pathogens by storing, preparing and using it correctly.

Information for industry or trade users should be clearly distinguishable from consumer information, particularly on food labels.

Lot Identification

Lot identification is essential in product recall and also helps effective stock rotation. Each container of food should be permanently marked to identify the producer and the lot. Codex General Standard for the Labelling of Prepackaged Foods applies.

Product Information

All food products should be accompanied by or bear adequate information to enable the next person in the food chain to handle, display, store and prepare and use the product safely and correctly.

Labelling

Prepackaged foods should be labelled with clear instructions to enable the next person in the food chain to handle, display store and use the product safely: Codex General Standard for the Labelling 1 of Prepackaged Foods applies.

Consumer Education

Health education programmes should enable consumers to understand the importance of any product information and to follow any instructions accompanying products, and make informed choices. In particular consumers should be informed of the relationship between time/temperature control and foodborne illness.

TRAINING

Training is fundamentally important to any food hygiene system. Inadequate hygiene training, and/or instruction and supervision of all people involved in food elated activities pose a potential threat to the safety of food and its suitability for consumption.

Food hygiene training is fundamentally important. All personnel should be aware of their role and responsibility in protecting food from contamination or deterioration. Food handlers should have the necessary knowledge and skills to enable them to handle food hygienically. Those who handle a strong cleaning chemicals or other potentially hazardous chemicals should be instructed in safe handling techniques.

Training Programmes

Factors to take into account in assessing the level of training required include:

— the nature of the food, in particular its ability to sustain growth of pathogenic or spoilage microorganisms;

— the manner in which the food is handled and packed, including the probability of contamination;

— the extent and nature of processing or further preparation before final consumption;

— the conditions under which the food will be stored; and

— the expected length of time before consumption.

— Training should be appropriate to the complexity of the manufacturing process and the tasks assigned.

— Personnel should be trained to understand the importance of the critical control points for which they are responsible, the critical limits, the procedures for monitoring, the action to be taken if the limits are not met and the records to be kept.

— Personnel responsible for maintenance of equipment having an impact on food safety should be appropriately trained to identify deficiencies that could affect product safety and to take appropriate corrective action, e.g. in-house repairs, contract repairs. Individuals performing maintenance on specific equipment, e.g. closing machines, recorders, etc., should be appropriately trained.

— Personnel and supervisors responsible for the sanitation programme should be appropriately trained to understand the principles and methods of effective cleaning and sanitizing.

— Additional training, e.g. specific technical training, apprenticeship programmes, etc., should be provided as necessary to ensure current knowledge of equipment and process technology.

Periodic assessments of the effectiveness of training and instruction programmes should be made, as well as routine supervision and checks to ensure that procedures are being carried out effectively. Managers and supervisors of food processes should have the necessary knowledge of food hygiene principles and practices to be able to judge potential risks and take the necessary action to remedy deficiencies.

Refresher Training

Training programmes should be routinely reviewed and updated where necessary. Systems should be in place to ensure that food handlers remain aware of all procedures necessary to maintain the safety and suitability of food. The manufacturer should have a written training programme for employees which should be delivered as follows:

— Appropriate training in personal hygiene and hygienic handling of food should be provided to all food handlers at the beginning of their employment.

— The original food hygiene training should be reinforced and updated at appropriate intervals.

The General Principles of Food Hygiene provide the basis for food hygiene and lay a firm foundation for the development of an effective HACCP or equivalent system. The application of the general principles and of good manufacturing practices (GMPs) allows the producer to operate within environmental conditions favourable to the production of safe food.

BIBLIOGRAPHY

Brown, L. J. "Risk/benefit-a rational and proactive approach to food safety." *The Nutrition Letter*, 8(3), PENpages No.09201263, July 1990.

Callahan, M. "Case of the Food-Safety Slip-up. (A play)". *Choices*, April 1991, 34-37.

Gantz, H., "Food safety program targets pre-teens". Food News for consumers, Summer 1990.

Gilmore, G., Meehan-Strub, M., and D. Mormann, "Community-based food safety survey", Journal of Extension, Spring 1992, 37-38.

Hall, R. L., "Food safety and biotechnology", *Nutrition Today*, May/ June 1991.

Hunter, L. J., and Branthaver, B. J. "Food Safety: A Nationwide Survey of Extension's Programs in Food Safety, *Planning, Development and Evaluation and Home Economics and Human Nutrition Extension Service*, USDA, 1990.

Morgan, B. L. G., "Food poisoning and traveller's diarrhea", *Nutrion and Health*, 9(2), September 1987.

Reynolds, S. "Food safety: Crockpot food safety", *Food and Nutriton News*, 12(3), March 1984.

Schaffner, D. "Understanding perceptions of food safety risks", *Nutrition News*, 1990, 53(2), 5-7.

Smith, S. M. "How to score a kitchen for health", *Environmental Nutrition*, 11(8), August 1988.

Street, R., "Safety first for summertime foods." *Better Homes and Garden*, August 1991.

Treu, J. A., "Food poisoning: Long-term side effects?" *Nutrition Letter*, 9(2), April 1991.

USDA, "The safe food book: your kitchen guide", *Home & Garden Bulletin No. 241.*. USDA Food Safety and Inspection Service, June 1985.

______________, "A Margin of Safety: The HACCP Approach to Food Safety Education Summary Report", *United States Department of Agriculture Food Safety and Inspection Service Information and Legislative Affairs,* 1989.

______________, "Play it safe: Goals for food safety", *Food safety inspection service,* 1990.

United Statetes Environmental Protection Agency, Environmental health risk education for youth: A resource manual. August 1991, A-107, 21K-1010.

INDEX